Solid Mechanics and Its Applications

Volume 200

Series Editor

G. M. L. Gladwell
Department of Civil Engineering, University of Waterloo, Waterloo, Canada

For further volumes:
http://www.springer.com/series/6557

Aims and Scope of the Series

The fundamental questions arising in mechanics are: *Why?*, *How?*, and *How much?* The aim of this series is to provide lucid accounts written by authoritative researchers giving vision and insight in answering these questions on the subject of mechanics as it relates to solids.

The scope of the series covers the entire spectrum of solid mechanics. Thus it includes the foundation of mechanics; variational formulations; computational mechanics; statics, kinematics and dynamics of rigid and elastic bodies: vibrations of solids and structures; dynamical systems and chaos; the theories of elasticity, plasticity and viscoelasticity; composite materials; rods, beams, shells and membranes; structural control and stability; soils, rocks and geomechanics; fracture; tribology; experimental mechanics; biomechanics and machine design.

The median level of presentation is the first year graduate student. Some texts are monographs defining the current state of the field; others are accessible to final year undergraduates; but essentially the emphasis is on readability and clarity.

Johan Blaauwendraad · Jeroen H. Hoefakker

Structural Shell Analysis

Understanding and Application

 Springer

Johan Blaauwendraad
Emeritus Professor
Delft University of Technology
Delft
The Netherlands

Jeroen H. Hoefakker
Former Lecturer
Delft University of Technology
Delft
The Netherlands

ISSN 0925-0042
ISBN 978-94-017-7723-0   ISBN 978-94-007-6701-0   (eBook)
DOI 10.1007/978-94-007-6701-0
Springer Dordrecht Heidelberg New York London

Printed on acid-free paper

Springer is part of Springer Science+Business Media (www.springer.com)

Preface

The structural and architectural potential of shell structures is used in various fields of civil, architectural, mechanical, aeronautical, and marine engineering. The strength of a (doubly) curved structure is efficiently and economically used, for example, to cover large areas without supporting columns. In addition to the mechanical advantages, the use of shell structures leads to esthetic architectural appearance. Examples of shells used in civil and architectural engineering include shell roofs, liquid storage tanks, silos, cooling towers, containment shells of nuclear power plants, and arch dams. Piping systems, curved panels, pressure vessels, bottles, buckets, and parts of cars, are examples of shells used in mechanical engineering. In aeronautical and marine engineering, shells are used in aircraft, spacecraft, missiles, ships, and submarines.

Similar to plate structures, one dimension of shell structures is small compared to the others. However, because of their spatial shape, the behavior of shells is different from that of plates. In flat plates, external loads are carried either by membrane response or bending response. In shells, the loads are carried by both. Similarly, both extensions and changes of curvature occur. As a result a mathematical description of the properties of a shell is much more elaborate than for beam and plate structures. Therefore many engineers and architects are unacquainted with aspects of shell behavior and design.

It took tens of years in the twentieth century to achieve sufficiently reliable shell theories for the different shell types that occur. Some of the most famous names in this respect are Love, Reissner, Wlassow, Morley, Flügge, Novoshilov, Koiter, Donnell, and Niordson. Well-known textbooks on the subject have also been published by Plüger, Rüdiger, Timoshenko, and Wolmir. Rather than contributing to theory development, this book is a university textbook, with a focus on architectural and civil engineering schools. Of course, practising professionals will profit from it as well. In writing this book we had three aims: (i) providing insight into the behavior of shell structures, (ii) explaining applied shell theories, and (iii) applying numerical programs for design purposes.

The book deals only with thin elastic shells, in particular with cylindrical, conical and spherical shells, and elliptic and hyperbolic paraboloids. The focus is on roofs, chimneys, pressure vessels, and storage tanks. The reader is supposed to be acquainted with the theory of flat plates loaded in-plane (shear walls, etc.) and loaded laterally (slabs, etc.). Material nonlinearity is not considered, and the deformation due to transverse shear is not taken into account. Geometric nonlinearity is considered only in an introductory chapter on buckling of thin shells. A substantial part of the book is derived from research efforts in the middle of the twentieth century at the Civil Engineering Department of Delft University of Technology by Bouma, Loof, and Van Koten. As such, we offer an addition to the archive of literature dealing with developments in shell research that are of continuing importance. Newer parts of the book come from doctoral thesis work of Hoefakker under supervision of Blaauwendraad [18].

The triple aim of the book is realized in the following way. We explain the theory of shells for a number of shell types. We show structural designers how to perform a manual calculation of the main force flow in a shell structure. We teach them how to estimate the stresses and the deformations. Special attention is paid to the characterization of edge bending effects. This is of prominent importance for mesh design in edge zones in case the structural designer performs a Finite Element Analysis.

Acknowledgments

Donnell's theory of shallow shells was the basis of extensive shell research in The Netherlands in the fifties and sixties of the 20th century. A team of people in the Civil Engineering Faculty of Delft University and the TNO research organisation for Building Research elaborated the theory, including work of Von Kármán and Jenkins. We acknowledge Professor A.L. Bouma, the late Associate Professor H.W. Loof and Mr. H. van Koten for their intensive academic effort to make the theory accessible for design purposes. A large part of the book has been adapted from their lecture notes on shell analysis in past decades, particularly the chapters on roof structures. We are indebted to Professor L.J. Sluys, editor-in-chief of HERON, joint publication of TNO, Delft University of Technology and Eindhoven University of Technology in The Netherlands. He permitted copying of tables and figures from bygone volumes. The extension on chimneys and tanks, on the basis of the Morley-Koiter theory, is more recent work of the authors, reflecting lectures on shell theory and doctoral thesis work. The persevering support of software virtuoso and structural engineer M.J. de Rijke in performing complex FEanalyses is highly rewarded. We warmly thank Springer ladies Nathalie Jacobs and Cynthia Feenstra for dedicated promotion and assistance respectively.

Contents

Part V Capita Selecta

Chapter 1
Introduction to Shells

This book is concerned with thin elastic shells. A *thin shell* has a very small thickness-to-minimal-radius ratio, often smaller than 1/50. As with plates, an applied load that acts out-of-plane leads to larger displacements than those generated by a load acting in-plane with the same intensity. Due to its initial curvature, a shell is able to transfer an applied load by in-plane as well as out-of-plane actions. A thin shell subjected to an applied load therefore produces mainly in-plane actions, which are called *membrane forces*. These membrane forces are actually resultants of normal stresses and in-plane shear stresses that are uniformly distributed across the thickness.

1.1 Shell Theories

A shell is a generalization of an isotropic homogeneous plate. Plates are flat structures of which the dimensions in two in-plane directions are large compared to the third direction perpendicular to the plate. The span in two directions is much larger than the thickness. Plates are defined by their *middle plane*, *thickness* and *material properties*. The *displacements* of the middle plane play the role of degrees of freedom in structural modelling. In-plane loads of plates generate in-plane membrane forces, and out-of-plane loads generate moments and transverse shear forces.

Shells are also defined by their *middle plane*, *thickness* and *material properties*. The difference with plates is that the middle plane of plates is flat, and that it is curved in shells. As a consequence, shells can carry out-of-plane loads by in-plane membrane forces, which is not possible for plates. In fact, this is the major reason why shells are such strong and economic structures.

The theory of this membrane behaviour is called *membrane theory*. However, membrane theory does not satisfy all equilibrium and/or displacement requirements of all cases. For example (Fig. 1.1):

J. Blaauwendraad and J. H. Hoefakker, *Structural Shell Analysis*,
Solid Mechanics and Its Applications 200, DOI: 10.1007/978-94-007-6701-0_1,
© Springer Science+Business Media Dordrecht 2014

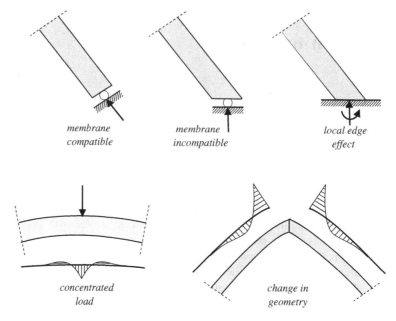

membrane   *membrane*   *local edge*
compatible   *incompatible*   *effect*

concentrated   *change in*
load   *geometry*

Fig. 1.1 Membrane and bending conditions

- Boundary conditions and deformation constraints that are incompatible with the requirements of a pure membrane field;
- Concentrated loads;
- Changes in geometry.

In the regions where the membrane theory will not hold, some (or all) of the bending field components are produced to compensate the shortcomings of the membrane field in the disturbed zone. These disturbances have to be described by a more complete analysis, which will lead to a *bending theory* of thin elastic shells.

If the bending components occur, they often have a local range of influence. Theoretical calculations and experiments show that the required bending components attenuate and often bending is confined to boundaries where a pure membrane solution does not exist. Therefore, in many cases the bending behaviour is restricted to an *edge disturbance*. The undisturbed and major part of the shell behaves like a true membrane. This unique property of shells is a result of the curvature of the spatial structure. Efficient structural performance is responsible for the widespread appearance of shells in nature. The continuous progress of numerical methods for computational mechanics, combined with an efficient structural performance and a pleasing shape, makes the application of shell structures more and more possible and favourable. However, for the use of numerical programs, some basic knowledge of the underlying theories and the mechanical behaviour of the structure is needed.

Many theories have been developed to analyse the mechanical behaviour of shell structures. To overcome the complexity of an exact theory, assumptions are made to produce simpler theories of which membrane theory is the most appealing. Because of its simplicity, membrane theory gives a direct insight into the structural behaviour and the order of magnitude of the expected response without elaborate computations. Membrane theory is thus very useful for initial design and analysis. Armed with a basic knowledge of the shell's more complex behaviour, we can take a first step towards a rational mechanical design. This book has been written to help readers develop insight into the design of shell structures.

1.2 Classification of Shells

1.2.1 Gaussian Curvature of a Surface

The geometry of a shell is completely described by the curved shape of the middle surface and the thickness distribution of the shell. At any point A on a smooth surface, there is a tangent plane. The normal to the surface at that point is defined to be the normal to the tangent plane, as shown in Fig. 1.2. A plane through point A that contains the normal is said to be *normal* to the middle surface at A. We call the intersections of the normal planes and the middle surface *normal sections* of the surface at A. Each of these curves has a local *curvature k* and a corresponding *radius of curvature r*. If the origin of the radius is at the positive side of the normal to the surface, the relation between the radius and the curvature is $r = k$. If the origin is at the negative side of the normal, the relation is $r = -k$.

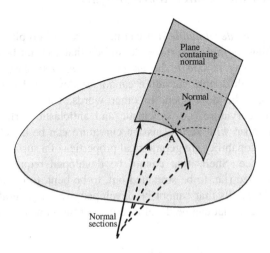

Fig. 1.2 Intersections of a plane with a surface

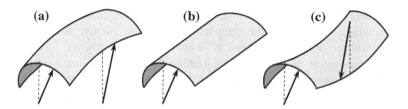

Fig. 1.3 Types of Gaussian curvature. **a** Positive Gaussian curvature. **b** Zero Gaussian curvature. **c** Negative Gaussian curvature

One of the infinitely many plane curves at point A has a minimum curvature, and another has a maximum curvature. These two plane curves are called the principal sections, and their curvatures, denoted by k_1 and k_2, are called the *principal curvatures* of the surface at A. By differential geometry, it can be proved that these two intersecting principal sections are orthogonal to each other. Because of this, it is convenient to take two axes of a local co-ordinate system on the surface along these principal sections. Taking a third axis normal to the surface at that point yields an orthogonal three-dimensional co-ordinate system.

The product of the two principal curvatures, $k_g = k_1 \cdot k_2$, is called the *Gaussian curvature* of the surface at A. The Gaussian curvature can be positive, negative, or zero. If it is positive, so that k_1 and k_2 have the *same* sign, the surface is said to be *synclastic* at that point. If it is negative, so that k_1 and k_2 have *opposite* signs, it is said to be *anticlastic*. If it is zero, one or both of k_1 and k_2 are zero. The surface is said to have a *single curvature*. The three types of surface are shown in Fig. 1.3.

1.2.2 Developed and Undeveloped Surfaces

A surface is said to be *developable* if it can be deformed into plane form without cutting or stretching its middle surface. A surface that can not be deformed into plane form in this way is said to be *undevelopable*. Both cases are shown in Fig. 1.4. Surfaces with double curvature cannot be developed, while those with single curvature can be developed. In other words, surfaces with positive or negative Gaussian curvature (i.e. synclastic and anticlastic surfaces) cannot be developed, while those with zero Gaussian curvature can be developed. Developability and undevelopability are geometrical properties of a surface, but they have structural significance. Shells that cannot be developed require more external energy to be deformed (i.e. to be stretched out, to be bent, to be destructed) than developable shells. Shells that cannot be developed are, in general, stronger and more stable than shells that can be developed having the same overall dimensions.

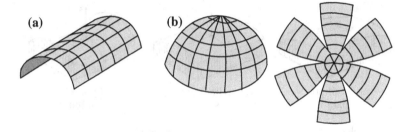

Fig. 1.4 Examples of a surface that can be developed (**a**), and a surface that cannot be developed (**b**)

1.2.3 Generated Surfaces

Surfaces of Revolution

Surfaces of revolution are generated by the revolution of a plane curve, called the meridional curve, about an axis, called the axis of revolution. In the special case of cylindrical and conical surfaces, the meridional curve consists of a line segment. Examples of surfaces of revolution are shown in Fig. 1.5.

Surfaces of Translation

Surfaces of translation are generated by sliding a plane curve along another plane curve, while keeping the orientation of the sliding curve constant. The curve on which the original curve slides, is called the *generator* of the surface. In the special case in which the generator is a straight line, the resulting surface is called a *cylindrical* surface. Examples of surfaces of translation are shown in Fig. 1.6.

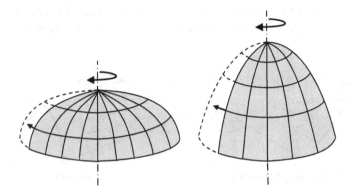

Fig. 1.5 Examples of surface of revolution

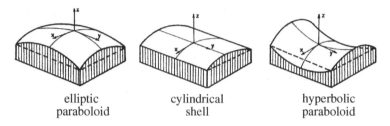

| elliptic | cylindrical | hyperbolic |
| paraboloid | shell | paraboloid |

Fig. 1.6 Examples of surfaces of translation with rectangular plan

Ruled Surfaces

Ruled surfaces are generated by sliding each end of a straight line on its own generator curve, while the straight line remains parallel to a prescribed plane. The sliding straight line is not necessarily at right angles to the planes containing the generator curves. From a practical point of view, moulding of on-site cast concrete shells having a ruled surface can be more easily and economically made by a rectilinear forming process. Examples of ruled surfaces are shown in Fig. 1.7.

1.2.4 Combined Surfaces

Combined surfaces can be partly synclastic and partly anticlastic, and can be composed of simpler shell forms. Combined surfaces can have discontinuous curvature. Examples of combined surfaces are shown in Fig. 1.8.

1.2.5 Folded Plates

A *folded plate* is a structure made up of two or more plates, as shown in Fig. 1.9. Strictly speaking, a folded plate is not a curved surface. Folded plates can be used to form very stiff three-dimensional structures. Several folded plate structures have

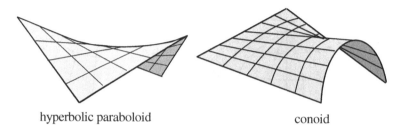

hyperbolic paraboloid   conoid

Fig. 1.7 Examples of ruled surfaces

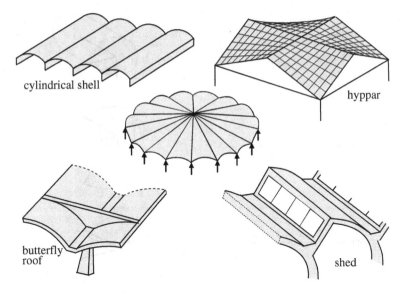

Fig. 1.8 Examples of combined surfaces

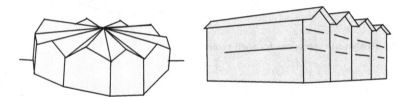

Fig. 1.9 Examples of folded plates

been made of rectangular, triangular or trapezium shaped plates. These structures can be analyzed only by numerical methods. Because of the high stiffness, large deformations are often prevented; the calculation can sometimes be made on the basis of a simple beam theory. Calculation methods for folded plate structures will not be treated in this book.

1.3 Analytical Description of the Shell Surface

1.3.1 Circular Plan

Shells with a circular plan include the cylinder, the cooling tower, the sphere and the ellipsoid. Analytical expressions for such shell shapes are presented in Fig. 1.10.

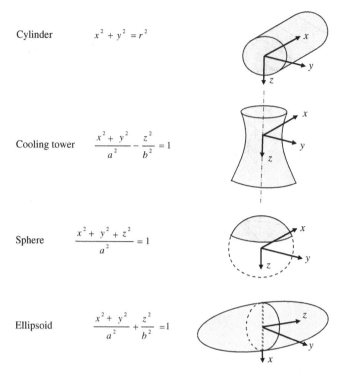

Cylinder $\qquad x^2 + y^2 = r^2$

Cooling tower $\qquad \dfrac{x^2 + y^2}{a^2} - \dfrac{z^2}{b^2} = 1$

Sphere $\qquad \dfrac{x^2 + y^2 + z^2}{a^2} = 1$

Ellipsoid $\qquad \dfrac{x^2 + y^2}{a^2} + \dfrac{z^2}{b^2} = 1$

Fig. 1.10 Analytical expressions for shell geometries

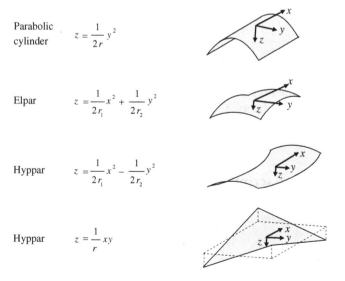

Parabolic
cylinder $\qquad z = \dfrac{1}{2r} y^2$

Elpar $\qquad z = \dfrac{1}{2r_1} x^2 + \dfrac{1}{2r_2} y^2$

Hyppar $\qquad z = \dfrac{1}{2r_1} x^2 - \dfrac{1}{2r_2} y^2$

Hyppar $\qquad z = \dfrac{1}{r} xy$

Fig. 1.11 Analytical expressions for shallow shell of rectangular plan

1.3.2 Rectangular Plan

In Fig. 1.11 examples are shown for shallow shells with a rectangular plan. In many cases a simple analytical expression of the form $z = f(x,y)$ can be used. In Fig. 1.11 this is done for a parabolic cylinder (as an approximation for the upper part of a cylindrical roof), an elliptic paraboloid (elpar), and two different descriptions of a hyperbolic paraboloid (hyppar).

Part I
Membrane Theory and
Edge Disturbances

Chapter 2
Membrane Theory for Shells with Principal Curvatures

The basic assumption of membrane theory is that a thin shell produces a pure membrane stress field, and that no bending stresses occur. This assumption is applicable if certain boundary and loading conditions, exemplified in Chap. 1, are met. In this pure membrane stress field, only normal and in-plane shear stresses are produced. They are due to stretching and shearing of the middle plane of the shell. Bending, torsion and transverse shear stresses are not accounted for.

We consider shells of arbitrary curvature and choose a set of axes x and y in the middle surface in the direction of the principal curvatures, see Fig. 2.1. The radius of curvature in x-direction is r_1 and the radius of curvature in y-direction is r_2. The z-axis is perpendicular to the middle surface. As drawn in this figure, the curvatures are negative, because the origin of the radius is at the negative normal side. The load may consist of three components: the in-plane components p_x and p_y, and the out-of-plane component p_z, in the directions x, y and z respectively. We define displacements u_x, u_y, and u_z of the middle surface, which correspond with the loads. Like in flat plates, normal stresses σ_{xx} and σ_{yy}, and a shear stress σ_{xy} occur. These are uniformly distributed through the thickness and integrate to n_{xx}, n_{yy}, and n_{xy}, respectively. The shear membrane force n_{yx} is equal to n_{xy} because of the moment equilibrium condition with respect to the normal axis z.

Notation and Sign Convention

The *notation* for stresses and forces in the membrane state may require clarification. For each stress or force we use two indices. The first indicates the face on which the stress or force acts. It is the direction of the normal of the face. The second index is the direction in which the stress is acting. For instance, the membrane force n_{xy} acts on a face with normal in the x-direction and is directed in the y-direction. The *sign convention* is as follows. A stress or force is *positive* if it acts in the *positive* coordinate direction on a plane with the normal vector in the *positive* coordinate direction. Correspondingly, a stress or force is positive if it acts in the

J. Blaauwendraad and J. H. Hoefakker, *Structural Shell Analysis*,
Solid Mechanics and Its Applications 200, DOI: 10.1007/978-94-007-6701-0_2,
© Springer Science+Business Media Dordrecht 2014

negative coordinate direction on a plane with its normal in the *negative* coordinate direction. As drawn in Fig. 2.1, the membrane forces are all positive.

The membrane forces, also called *stress resultants*, generate corresponding strains ε_{xx}, ε_{yy}, and γ_{xy}. Of these, the first two are normal strains, and the latter is a shear angle. It is convenient to define four vectors

$$\mathbf{u} = \left[u_x, u_y, u_z\right]^{\mathrm{T}} \tag{2.1}$$

$$\mathbf{p} = \left[p_x, p_y, p_z\right]^{\mathrm{T}} \tag{2.2}$$

$$\mathbf{e} = \left[\varepsilon_{xx}, \varepsilon_{yy}, \gamma_{xy}\right]^{\mathrm{T}} \tag{2.3}$$

$$\mathbf{s} = \left[n_{xx}, n_{yy}, n_{xy}\right]^{\mathrm{T}} \tag{2.4}$$

The four vectors are related to each other by three basic relationships: *kinematic relation*, *constitutive relation* and *equilibrium relation*. The kinematic relations relate strains to displacements, the constitutive relations relate membrane forces to strains, and equilibrium relations relate membrane forces to load components. The scheme of relationships is shown in Fig. 2.2.

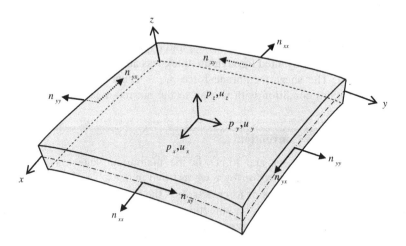

Fig. 2.1 Definition of displacements, loading and membrane forces. Sign convention

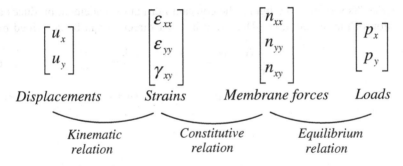

$$\begin{bmatrix} u_x \\ u_y \end{bmatrix} \qquad \begin{bmatrix} \varepsilon_{xx} \\ \varepsilon_{yy} \\ \gamma_{xy} \end{bmatrix} \qquad \begin{bmatrix} n_{xx} \\ n_{yy} \\ n_{xy} \end{bmatrix} \qquad \begin{bmatrix} p_x \\ p_y \end{bmatrix}$$

Displacements   *Strains*   *Membrane forces*   *Loads*

Kinematic relation   *Constitutive relation*   *Equilibrium relation*

Fig. 2.2 Scheme of relationships in membrane theory

2.1 Kinematic Relation

Recall that the displacements u_x and u_y are displacements tangential to the middle surface in the direction of the principal curvatures k_x and k_y respectively, and the displacement u_z is the displacement normal to the middle surface, in the direction of the z-axis. The description of the strain vector **e** due to the tangential displacements u_x and u_y is the same as the description for a plate loaded in its plane, which is:

$$\varepsilon_{xx} = \frac{\partial u_x}{\partial x}$$
$$\varepsilon_{yy} = \frac{\partial u_y}{\partial y} \qquad (2.5)$$
$$\gamma_{xy} = \frac{\partial u_x}{\partial y} + \frac{\partial u_y}{\partial x}$$

The third displacement, the normal displacement u_z, also contributes to the kinematic relations. The contribution of a constant normal displacement to the elongation in x-direction is shown in Fig. 2.3. The arc length of this strip is

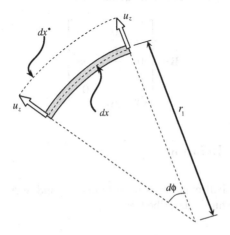

Fig. 2.3 Elongation due to the normal displacement u_z

$dx = r_1 d\phi$. As shown in the figure, the constant normal displacement produces an elongation of the middle fibre. The strain in the x-direction can be described by

$$\varepsilon_{xx} = \frac{dx^* - dx}{dx} = \frac{(r_1 + u_z)d\phi - r_1 d\phi}{r_1 d\phi} = \frac{u_z}{r_1} \tag{2.6}$$

The strain in the x-direction of the middle surface (with negative curvature) is thus equal to

$$\varepsilon_{xx} = \frac{u_z}{r_1} = -k_1 u_z \tag{2.7}$$

The same reasoning holds for the y-direction yielding the additional strain:

$$\varepsilon_{yy} = \frac{u_z}{r_2} = -k_2 u_z \tag{2.8}$$

The normal displacement does not alter the shear strain γ_{xy} because the co-ordinate system is placed in the direction of the principal curvatures.

Incorporating the additional normal strains of Eqs. (2.7) and (2.8) into Eq. (2.5) yields the complete kinematic relation for the membrane strains:

$$\begin{bmatrix} \varepsilon_{xx} \\ \varepsilon_{yy} \\ \gamma_{xy} \end{bmatrix} = \begin{bmatrix} \frac{\partial}{\partial x} & 0 & -k_1 \\ 0 & \frac{\partial}{\partial y} & -k_2 \\ \frac{\partial}{\partial y} & \frac{\partial}{\partial x} & 0 \end{bmatrix} \begin{bmatrix} u_x \\ u_y \\ u_z \end{bmatrix} \tag{2.9}$$

Symbolically, this kinematic relation can be written as:

$$\mathbf{e} = \mathbf{B}\mathbf{u} \tag{2.10}$$

The vectors $\mathbf{e}$ and $\mathbf{u}$ are defined by Eqs. (2.3) and (2.1) respectively, and the differential operator matrix $\mathbf{B}$ is thus defined as:

$$\mathbf{B} = \begin{bmatrix} \frac{\partial}{\partial x} & 0 & -k_1 \\ 0 & \frac{\partial}{\partial y} & -k_2 \\ \frac{\partial}{\partial y} & \frac{\partial}{\partial x} & 0 \end{bmatrix} \tag{2.11}$$

2.2 Constitutive Relation

By assumption, the shell material is linearly-elastic, and obeys Hooke's law. The stress strain relationship is described by:

$$\begin{bmatrix} \sigma_{xx} \\ \sigma_{yy} \\ \sigma_{xy} \end{bmatrix} = \frac{E}{1 - v^2} \begin{bmatrix} 1 & v & 0 \\ v & 1 & 0 \\ 0 & 0 & \frac{1-v}{2} \end{bmatrix} \begin{bmatrix} \varepsilon_{xx} \\ \varepsilon_{yy} \\ \gamma_{xy} \end{bmatrix} \tag{2.12}$$

The elasticity and the lateral contraction of the material are herein expressed by Young's modulus E and Poisson's ratio v respectively. Equation (2.12) yields the constitutive relation between the membrane forces (stress resultants) and the strains:

$$\begin{bmatrix} n_{xx} \\ n_{yy} \\ n_{xy} \end{bmatrix} = \frac{Et}{1 - v^2} \begin{bmatrix} 1 & v & 0 \\ v & 1 & 0 \\ 0 & 0 & \frac{1-v}{2} \end{bmatrix} \begin{bmatrix} \varepsilon_{xx} \\ \varepsilon_{yy} \\ \gamma_{xy} \end{bmatrix} \tag{2.13}$$

Symbolically, this stiffness relation can be written as

$$s = D e \tag{2.14}$$

where D is the *rigidity matrix*. This matrix is

$$D = \frac{Et}{1 - v^2} \begin{bmatrix} 1 & v & 0 \\ v & 1 & 0 \\ 0 & 0 & \frac{1-v}{2} \end{bmatrix} \tag{2.15}$$

2.3 Equilibrium Relation

From the scheme in Fig. 2.2 we know that three equilibrium equations must be derived, because we have three degrees of freedom with three external load components. The two equilibrium equations for the load components p_x and p_y tangential to the middle surface are the same as for a flat plate loaded in-plane:

$$\frac{\partial n_{xx}}{\partial x} + \frac{\partial n_{yx}}{\partial y} + p_x = 0$$
$$\frac{\partial n_{xy}}{\partial x} + \frac{\partial n_{yy}}{\partial y} + p_y = 0 \tag{2.16}$$

To obtain the third equilibrium equation for the load component p_z normal to the middle surface, the curvature of the middle surface needs to be investigated. Recall that the co-ordinate system is placed according to the principal curvatures, which are denoted by k_1 and k_2, and the curvatures are the reciprocals of the radii of curvature, which are r_1 and r_2 in the principal directions respectively. Furthermore, the curvature is positive or negative according to whether the corresponding centre of curvature lies on the positive or the negative side of the normal of the middle surface. The normal load component in the z-direction on a strip in the x-direction is shown in Fig. 2.4. This strip has a unit width and an arc length

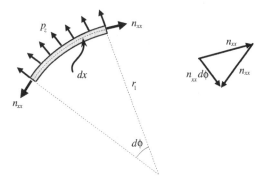

Fig. 2.4 Membrane force n_{xx} and normal load p_z

$dx = r_1 d\phi$, in which $d\phi$ is the infinitesimal opening angle in the x-direction. The total load on the strip under consideration is thus equal to $p_z \cdot r_1 d\phi$. The stress resultants n_{xx} at both ends of the strip are on an infinitesimal angle $d\phi$ to each other and their resultant in the negative z-direction is therefore equal to $n_{xx} d\phi$ as shown in the right-hand part of Fig. 2.4. Hence, the equilibrium condition for this strip yields the equation: $-n_{xx} \cdot d\phi + p_z \cdot r_1 d\phi = 0$. If we divide by $dx = r_1 d\phi$ and use the relationship $k_1 = -1/r_1$, we obtain $k_1 n_{xx} + p_z = 0$. A similar reasoning can be applied to a strip of unit width in the y-direction, leading to a positive contribution of $k_2 n_{yy}$ in the z-direction.

We now consider an elementary element as shown in Fig. 2.5. The length in x-direction is dx and in the y-direction dy, where $dx = r_1 d\phi$ and $dy = r_2 d\theta$. The load acts on the area $r_1 d\phi \cdot r_2 d\theta$, the membrane force n_{xx} acts on a face with arc length $r_2 d\theta$, and the stress resultant n_{yy} acts on a face with arc length $r_1 d\phi$. The equilibrium equation in the z-direction is therefore

$$-n_{xx} r_2 d\theta \cdot d\phi - n_{yy} r_1 d\phi \cdot d\theta + p_z \cdot r_1 d\phi \cdot r_2 d\theta = 0 \qquad (2.17)$$

We divide by the area $r_1 d\phi \cdot r_2 d\theta$ obtaining the third equilibrium equation for the membrane behaviour of a shell element with arbitrary curvature:

$$k_1 n_{xx} + k_2 n_{yy} + p_z = 0 \qquad (2.18)$$

All the equilibrium equations of the membrane theory for thin shells are now known. Eqs. (2.16) and (2.18) describe the equilibrium between the three load terms p_x, p_y, p_z and the three membrane forces n_{xx}, n_{yy}, n_{xy}:

$$\frac{\partial n_{xx}}{\partial x} + \frac{\partial n_{yx}}{\partial y} + p_x = 0$$

$$\frac{\partial n_{xy}}{\partial x} + \frac{\partial n_{yy}}{\partial y} + p_y = 0 \qquad (2.19)$$

$$k_1 n_{xx} + k_2 n_{yy} + p_z = 0$$

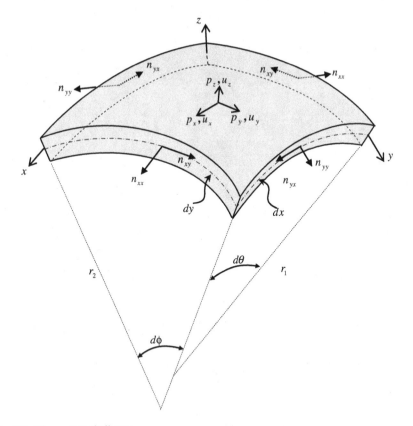

Fig. 2.5 Elementary shell part

This equilibrium relation is presented symbolically as

$$\mathbf{B}^*\mathbf{s} = \mathbf{p} \qquad (2.20)$$

The *differential operator matrix* $\mathbf{B}^*$ is

$$\mathbf{B}^* = \begin{bmatrix} -\dfrac{\partial}{\partial x} & 0 & -\dfrac{\partial}{\partial y} \\[2mm] 0 & -\dfrac{\partial}{\partial y} & -\dfrac{\partial}{\partial x} \\[2mm] -k_1 & -k_2 & 0 \end{bmatrix} \qquad (2.21)$$

$\mathbf{B}^*$ is the adjoint of the differential operator matrix $\mathbf{B}$ of Eq. (2.11) which appeared in the kinematic relations. $\mathbf{B}^*$ is mirrored with respect to the main diagonal of $\mathbf{B}$, and the derivatives have changed sign.

Chapter 3
Membrane Theory for Thin Shells of Arbitrary Curvatures

The three sets of equations of Chap. 2 derived for the membrane behaviour of thin shells relate to a co-ordinate system placed according to the principal curvatures. In practice it may be useful to choose the co-ordinate system in such a way that a co-ordinate axis is placed along an edge of the shell, which does not necessarily coincide with a principal curvature. This is the subject of this chapter. Different from the approach of Chap. 2, we now choose a reference system of axes in the tangent plane of a point O at the middle surface of the shell. The x-axis and y-axis are in the tangent plane, and the z-axis is normal to the plane. Instead of the principal curvatures k_1 and k_2, we now work with curvatures k_x and k_y. On top of that, it will appear to be convenient to define a twist k_{xy}. The expressions for the curvatures k_x and k_y are in fact the same as in Chap. 2, but we will derive them again in an alternate way, such that it is easy to extrapolate to the derivation of the twist k_{xy}.

3.1 Kinematic Relations

The description of the strain vector **e** due to the tangential displacements u_x and u_y is again the same as the description for a plate loaded in its plane, similar to Chap. 2:

$$\varepsilon_{xx} = \frac{\partial u_x}{\partial x}, \quad \varepsilon_{yy} = \frac{\partial u_y}{\partial y}, \quad \gamma_{xy} = \frac{\partial u_x}{\partial y} + \frac{\partial u_y}{\partial x} \tag{3.1}$$

For the effect of the normal displacement u_z, we consider the intersection line of the normal plane through the x-axis and the shell. In Fig. 3.1 we show an infinitesimal shell element of unit width and length $2dx$. The x-axis is in the tangent plane at O. After displacement over the distance u_z in the positive z-direction, the length dx increases to $dx + de_x$. We define the inclination $\varphi_x = -\partial z/\partial x$, which is zero in the origin at O. The minus sign is introduced because in Fig. 3.1 the coordinate z decreases with increasing x. As a consequence, the incremental change of the inclination over the distance dx is

J. Blaauwendraad and J. H. Hoefakker, *Structural Shell Analysis*,
Solid Mechanics and Its Applications 200, DOI: 10.1007/978-94-007-6701-0_3,
© Springer Science+Business Media Dordrecht 2014

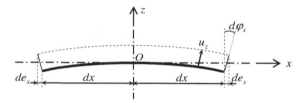

Fig. 3.1 Effect of displacement u_z in curved shell surface

$$d\varphi_x = -\frac{\partial^2 z}{\partial x^2} dx \tag{3.2}$$

The change of length is $de_x = u_z d\varphi_x$ from which we derive

$$de_x = -\frac{\partial^2 z}{\partial x^2} dx \cdot u_z \tag{3.3}$$

The second derivative in Fig. 3.1 is negative, so a positive elongation de_x occurs. The strain in the middle surface is

$$\varepsilon_{xx} = \frac{de_x}{dx} = -\frac{\partial^2 z}{\partial x^2} u_z \tag{3.4}$$

which we write as

$$\varepsilon_{xx} = -k_x u_z \tag{3.5}$$

where the curvature k_x is

$$k_x = \frac{\partial^2 z}{\partial x^2} \tag{3.6}$$

We obtained Eq. (3.5) earlier as Eq. (2.7), and we now learn that the curvature is equal to the second derivative of the middle surface, expressed in the coordinates of the tangent plane. Without further proof, we state that it holds similarly for the y-direction

$$\varepsilon_{yy} = -k_y u_z \tag{3.7}$$

where

$$k_y = \frac{\partial^2 z}{\partial y^2} \tag{3.8}$$

and

$$d\varphi_x = -\frac{\partial^2 z}{\partial x^2} dx \tag{3.9}$$

Next we study the infinitesimal twisted shell part as drawn in Fig. 3.2. The dotted lines represent the tangent plane and the full lines are the shell of twisted

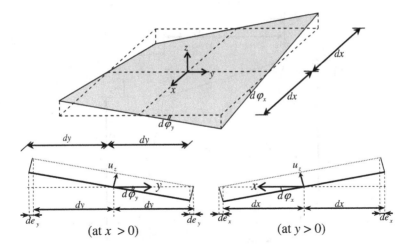

Fig. 3.2 Effect of displacement u_z in twisted shell surface

shape. We introduce inclinations $\varphi_x = -\partial z/\partial x$ and $\varphi_y = -\partial z/\partial y$. The incremental changes $d\varphi_x$ in y-direction and $d\varphi_y$ in x-direction are

$$d\varphi_x = -\frac{\partial^2 z}{\partial y \partial x}\,dy, \ d\varphi_y = -\frac{\partial^2 z}{\partial x \partial y}\,dx \tag{3.10}$$

For a continuous surface such as we are studying, the two mixed second derivatives in this equation are equal to each other. In the bottom part of Fig. 3.2 we show the displacement of an edge of the shell part over a distance u_z in z-direction. Both ends of the edge displace over the distance de_x in the negative x-direction. At the opposite edge displacements de_x will occur in the positive direction. Similarly, displacements de_y occur along the other two edges. Figure 3.3 shows the displacements of all four corners of the element. Note that the rectangular middle surface has deformed into a rhombus, so a twist in the middle surface generates a shear strain γ_{xy} in case of a normal displacement u_z. The size of the shear strain is derived as follows. From Fig. 3.2 we derive

$$de_x = d\varphi_x\,u_z, \ de_y = d\varphi_y\,u_z \tag{3.11}$$

and from Fig. 3.3,

$$\varepsilon_{xy} = \frac{de_x}{dy}, \ \varepsilon_{yx} = \frac{de_y}{dx} \tag{3.12}$$

We obtain the expression for the strain $\gamma_{xy} = \varepsilon_{xy} + \varepsilon_{yx}$. Accounting for Eqs. (3.10), (3.11) and (3.12) we find that ε_{xy} and ε_{yx} are equal and obtain

$$\gamma_{xy} = -2k_{xy}\,u_z \tag{3.13}$$

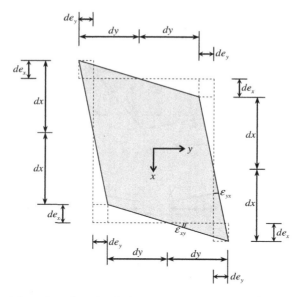

Fig. 3.3 Shear deformation of a twisted middle surface due to a normal displacement u_z

where k_{xy} is defined by the mixed second curvature of the middle surface

$$k_{xy} = \frac{\partial^2 z}{\partial x \partial y} \tag{3.14}$$

Equations (3.1), (3.5), (3.7) and (3.13) jointly yield the kinematic relation:

$$\begin{bmatrix} \varepsilon_{xx} \\ \varepsilon_{yy} \\ \gamma_{xy} \end{bmatrix} = \begin{bmatrix} \dfrac{\partial}{\partial x} & 0 & -k_x \\ 0 & \dfrac{\partial}{\partial y} & -k_y \\ \dfrac{\partial}{\partial y} & \dfrac{\partial}{\partial x} & -2k_{xy} \end{bmatrix} \begin{bmatrix} u_x \\ u_y \\ u_z \end{bmatrix} \tag{3.15}$$

symbolically presented as $\mathbf{e} = \mathbf{Bu}$. The differential operator matrix $\mathbf{B}$ is

$$\mathbf{B} = \begin{bmatrix} \dfrac{\partial}{\partial x} & 0 & -k_x \\ 0 & \dfrac{\partial}{\partial y} & -k_y \\ \dfrac{\partial}{\partial y} & \dfrac{\partial}{\partial x} & -2k_{xy} \end{bmatrix} \tag{3.16}$$

and the three curvatures are defined by

$$k_x = \frac{\partial^2 z}{\partial x^2}, \; k_y = \frac{\partial^2 z}{\partial y^2}, \; k_{xy} = \frac{\partial^2 z}{\partial x \partial y} \tag{3.17}$$

In case the x- and y-axis are chosen in the direction of the principal curvatures, the twist k_{xy} will be zero and Eq. (3.15) is equal to Eq. (2.9).

3.2 Constitutive Equation

Equations (2.13), (2.14) and (2.15) remain unchanged for arbitrary curvatures. They are reproduced here:

$$\begin{bmatrix} n_{xx} \\ n_{yy} \\ n_{xy} \end{bmatrix} = \frac{Et}{1-v^2} \begin{bmatrix} 1 & v & 0 \\ v & 1 & 0 \\ 0 & 0 & \frac{1-v}{2} \end{bmatrix} \begin{bmatrix} \varepsilon_{xx} \\ \varepsilon_{yy} \\ \gamma_{xy} \end{bmatrix} \tag{3.18}$$

Symbolically, we write again this stiffness relation as

$$\mathbf{s} = \mathbf{D}\,\mathbf{e} \tag{3.19}$$

where $\mathbf{D}$ is the *rigidity matrix*:

$$\mathbf{D} = \frac{Et}{1-v^2} \begin{bmatrix} 1 & v & 0 \\ v & 1 & 0 \\ 0 & 0 & \frac{1-v}{2} \end{bmatrix} \tag{3.20}$$

3.3 Equilibrium Equation

Compared to Sect. 2.3 the in-plane equilibrium equations in the now considered co-ordinate system do not change:

$$\begin{aligned} \frac{\partial n_{xx}}{\partial x} + \frac{\partial n_{xy}}{\partial y} + p_x &= 0 \\ \frac{\partial n_{xy}}{\partial x} + \frac{\partial n_{yy}}{\partial y} + p_y &= 0 \end{aligned} \tag{3.21}$$

The out-of-plane equilibrium equation is derived in two steps. First the consequence of the two curvatures k_x and k_y is derived, next the effect of the twist k_{xy}.

3.3.1 Effect of Curvatures

For the first step we consider Fig. 3.4 with a shell strip of unit width and length dx. The shell part covers an angle $d\varphi_x$. The downward resultant of the membrane

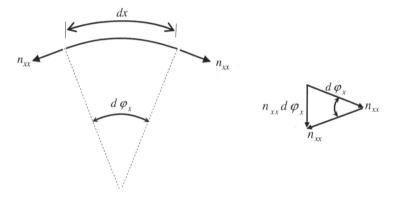

Fig. 3.4 Downward resultant of membrane force n_{xx} over length dx

forces n_{xx} on the left-hand and right-hand ends is $n_{xx}d\varphi_x$, and the total upward distributed load over this shell strip is $p_z dx$. This leads to the equilibrium equation per unit width

$$p_z dx - n_{xx}d\varphi_x = 0 \tag{3.22}$$

Accounting for Eqs. (3.2) and (3.6) we obtain

$$k_x n_{xx} + p_z = 0 \tag{3.23}$$

Similarly we may consider a shell strip in y-direction, which has unit width and length dy. Now the downward resultant of n_{yy} is $n_{yy}d\varphi_y$, and the equilibrium equation becomes

$$p_z dy - n_{yy}d\varphi_y = 0 \tag{3.24}$$

which leads to

$$k_y n_{yy} + p_z = 0 \tag{3.25}$$

We combine Eqs. (3.23) and (3.25) by considering a shell part with sizes dx in x-direction and dy in y-direction, respectively. The downward resultant $n_{xx}d\varphi_x$ acts over the width dy, the downward resultant $n_{yy}d\varphi_y$ over the width dx, and the load p over the area $dxdy$. Hence, equilibrium requires

$$p_z dxdy - n_{xx}d\varphi_x dy - n_{yy}d\varphi_y dx = 0 \tag{3.26}$$

Accounting for Eqs. (3.2), (3.6), (3.8) and (3.9), we obtain

$$k_x n_{xx} + k_y n_{yy} + p_z = 0 \tag{3.27}$$

which is comparable with Eq. (2.18) in Chap. 2.

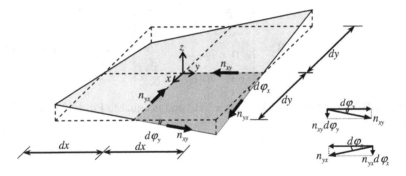

Fig. 3.5 Contribution of the shear membrane forces to equilibrium in z-direction

3.3.2 Effect of Twist

Now we execute the second step, the derivation of the contribution of the twist k_{xy}, for which we refer the reader to Fig. 3.5. The shear membrane force n_{xy} acts in the horizontal edge $x = 0$ and the inclined edge at a distance dx. These two forces produce a vertical downward resultant $n_{xy}d\varphi_y$, which acts over a width dx. Similarly a vertical downward resultant $n_{yx}d\varphi_x$ occurs over a width dy. The equilibrium equation in z-direction is

$$p_z\,dx\,dy - (n_{xy}d\varphi_y)dx - (n_{yx}d\varphi_x)dy = 0 \qquad (3.28)$$

Substituting Eq. (3.10) and accounting for Eq. (3.14), this equation becomes (dividing by $dxdy$)

$$2k_{xy}n_{xy} + p_z = 0 \qquad (3.29)$$

We obtain the final equilibrium relation by assembling the Eqs. (3.21), (3.27) and (3.29):

$$\begin{bmatrix} -\dfrac{\partial}{\partial x} & 0 & -\dfrac{\partial}{\partial y} \\[2mm] 0 & -\dfrac{\partial}{\partial y} & -\dfrac{\partial}{\partial x} \\[2mm] -k_x & -k_y & -2k_{xy} \end{bmatrix} \begin{bmatrix} n_{xx} \\ n_{yy} \\ n_{xy} \end{bmatrix} = \begin{bmatrix} p_x \\ p_y \\ p_z \end{bmatrix} \qquad (3.30)$$

Symbolically we write again

$$\mathbf{s} = \mathbf{B}^*\mathbf{p} \qquad (3.31)$$

where the differential operator matrix $\mathbf{B}^*$ is

$$\mathbf{B}^* = \begin{bmatrix} -\dfrac{\partial}{\partial x} & 0 & -\dfrac{\partial}{\partial y} \\ 0 & -\dfrac{\partial}{\partial y} & -\dfrac{\partial}{\partial x} \\ -k_x & -k_y & -2k_{xy} \end{bmatrix} \qquad (3.32)$$

Matrix $\mathbf{B}^*$ is the adjoint of $\mathbf{B}$ in Eq. (3.16) and is equal to Eq. (2.23) if the twist k_{xy} is zero.

Chapter 4
Application of Membrane Theory to Circular Cylindrical Shells

4.1 Description of the Circular Cylindrical Surface

For a circular cylindrical shell, it is convenient to apply a polar co-ordinate system to the cross-sectional profile as illustrated in Fig. 4.1. The axes are chosen in the longitudinal, circumferential and transverse directions. For the circumferential ordinate y on the middle surface the equality $y = a\theta$ holds, in which the constant radius of the circular cylinder is denoted by a. Instead of k_y, p_y and u_y, we will write k_θ, p_θ and u_θ, and we replace n_{yy} and n_{xy} by $n_{\theta\theta}$ and $n_{x\theta}$. An infinitesimal element has thus sides with length, measured on the middle surface, dx in the longitudinal and $dy = a\,d\theta$ in the circumferential direction. The thickness of the element is denoted by t. The three positive directions of the displacements (u_x, u_θ, u_z) are taken correspondingly to the three positive directions of the co-ordinates (x, θ, z). The straight generatrix in x-direction has a curvature $k_x = 0$ and therefore a radius of curvature $r_x = \infty$. The radius of curvature in the circumferential direction is $r_\theta = a = constant$. The corresponding curvature is $k_\theta = -1/a$. Since the co-ordinate axes are placed according to the principal curvatures, the twist $k_{x\theta}$ is zero. Hereafter we will refer to the membrane theory of Chap. 3. The following vectors are used:

$$\begin{aligned}
\mathbf{u} &= \begin{bmatrix} u_x & u_\theta & u_z \end{bmatrix}^T \\
\mathbf{e} &= \begin{bmatrix} \varepsilon_{xx} & \varepsilon_{\theta\theta} & \gamma_{x\theta} \end{bmatrix}^T \\
\mathbf{s} &= \begin{bmatrix} n_{xx} & n_{\theta\theta} & n_{x\theta} \end{bmatrix}^T \\
\mathbf{p} &= \begin{bmatrix} p_x & p_\theta & p_z \end{bmatrix}^T
\end{aligned} \qquad (4.1)$$

The relation between these vectors is presented in Fig. 4.2.

J. Blaauwendraad and J. H. Hoefakker, *Structural Shell Analysis*,
Solid Mechanics and Its Applications 200, DOI: 10.1007/978-94-007-6701-0_4,
© Springer Science+Business Media Dordrecht 2014

4.2 Kinematic Relation

For the circular cylindrical shell the kinematic relation (3.15) becomes

$$\begin{bmatrix} \varepsilon_{xx} \\ \varepsilon_{\theta\theta} \\ \gamma_{x\theta} \end{bmatrix} = \begin{bmatrix} \dfrac{\partial}{\partial x} & 0 & 0 \\ 0 & \dfrac{1}{a}\dfrac{\partial}{\partial \theta} & \dfrac{1}{a} \\ \dfrac{1}{a}\dfrac{\partial}{\partial \theta} & \dfrac{\partial}{\partial x} & 0 \end{bmatrix} \begin{bmatrix} u_x \\ u_\theta \\ u_z \end{bmatrix} \tag{4.2}$$

which is symbolically presented as $\mathbf{e} = \mathbf{B}\mathbf{u}$. The matrix $\mathbf{B}$ is

$$\mathbf{B} = \begin{bmatrix} \dfrac{\partial}{\partial x} & 0 & 0 \\ 0 & \dfrac{1}{a}\dfrac{\partial}{\partial \theta} & \dfrac{1}{a} \\ \dfrac{1}{a}\dfrac{\partial}{\partial \theta} & \dfrac{\partial}{\partial x} & 0 \end{bmatrix} \tag{4.3}$$

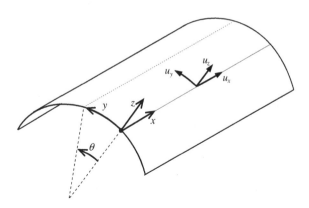

Fig. 4.1 Co-ordinate system and corresponding displacements of the middle surface

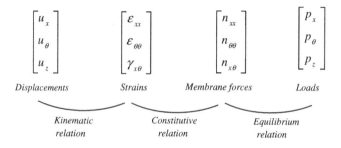

$$\begin{bmatrix} u_x \\ u_\theta \\ u_z \end{bmatrix} \qquad \begin{bmatrix} \varepsilon_{xx} \\ \varepsilon_{\theta\theta} \\ \gamma_{x\theta} \end{bmatrix} \qquad \begin{bmatrix} n_{xx} \\ n_{\theta\theta} \\ n_{x\theta} \end{bmatrix} \qquad \begin{bmatrix} p_x \\ p_\theta \\ p_z \end{bmatrix}$$

Displacements   *Strains*   *Membrane forces*   *Loads*

Kinematic   *Constitutive*   *Equilibrium*
relation   *relation*   *relation*

Fig. 4.2 Scheme of relationships

4.3 Constitutive Relation

The constitutive equation (3.18) is rewritten as

$$\begin{bmatrix} n_{xx} \\ n_{\theta\theta} \\ n_{x\theta} \end{bmatrix} = \frac{Et}{1-\nu^2} \begin{bmatrix} 1 & \nu & 0 \\ \nu & 1 & 0 \\ 0 & 0 & \frac{1-\nu}{2} \end{bmatrix} \begin{bmatrix} \varepsilon_{xx} \\ \varepsilon_{\theta\theta} \\ \gamma_{x\theta} \end{bmatrix} \tag{4.4}$$

which is symbolically presented as

$$\mathbf{s} = \mathbf{D}\mathbf{e} \tag{4.5}$$

The rigidity matrix $\mathbf{D}$ is defined by

$$\mathbf{D} = \frac{Et}{1-\nu^2} \begin{bmatrix} 1 & \nu & 0 \\ \nu & 1 & 0 \\ 0 & 0 & \frac{1-\nu}{2} \end{bmatrix} \tag{4.6}$$

4.4 Equilibrium Relation

The equilibrium equation (3.30) is rewritten as

$$\begin{bmatrix} -\dfrac{\partial}{\partial x} & 0 & -\dfrac{1}{a}\dfrac{\partial}{\partial\theta} \\ 0 & -\dfrac{1}{a}\dfrac{\partial}{\partial\theta} & -\dfrac{\partial}{\partial x} \\ 0 & \dfrac{1}{a} & 0 \end{bmatrix} \begin{bmatrix} n_{xx} \\ n_{\theta\theta} \\ n_{x\theta} \end{bmatrix} = \begin{bmatrix} p_x \\ p_\theta \\ p_z \end{bmatrix} \tag{4.7}$$

which is symbolically presented as $\mathbf{B}^*\mathbf{s} = \mathbf{p}$. The matrix $\mathbf{B}^*$ is

$$\mathbf{B}^* = \begin{bmatrix} -\dfrac{\partial}{\partial x} & 0 & -\dfrac{1}{a}\dfrac{\partial}{\partial\theta} \\ 0 & -\dfrac{1}{a}\dfrac{\partial}{\partial\theta} & -\dfrac{\partial}{\partial x} \\ 0 & \dfrac{1}{a} & 0 \end{bmatrix} \tag{4.8}$$

4.5 General Solution

The set of three equilibrium Eq. (4.7) with three unknowns is statically determined. Moreover, since the membrane force in circumferential direction $n_{\theta\theta}$ is directly known from the third equation for the equilibrium in normal direction, the equations can be consecutively solved by a separate equilibrium equation. The general solution for the membrane forces thus becomes

$$n_{\theta\theta} = ap_z$$

$$n_{x\theta} = -\int \left(p_\theta + \frac{1}{a} \frac{\partial n_{\theta\theta}}{\partial\theta} \right) dx$$

$$n_{xx} = -\int \left(p_x + \frac{1}{a} \frac{\partial n_{x\theta}}{\partial\theta} \right) dx \tag{4.9}$$

The integrations yield two unknown functions. These functions contain two constants that are computed by the boundary conditions under consideration. However, the membrane force $n_{\theta\theta}$ depends only on the local intensity of the normal load p_z and cannot be influenced by boundary conditions. This feature of the membrane theory is not of great importance for shells whose cross-sectional profiles are closed curves and which have only two cross-sectional profiles as boundaries. Hereby, the constants must be determined from two boundary conditions, which must be of a kind that on a profile $x = constant$ (one end of the shell or a plane of symmetry) one of the membrane forces $n_{x\theta}$ or n_{xx} is given as an arbitrary function of the ordinate θ.

Once the vector $\mathbf{n}$ of membrane forces is known, the constitutive Eq. (4.4) are inverted to arrive at an expression for the deformation vector $\mathbf{e} = \mathbf{D}^{-1}\mathbf{s}$:

$$\begin{bmatrix} \varepsilon_{xx} \\ \varepsilon_{\theta\theta} \\ \gamma_{x\theta} \end{bmatrix} = \frac{1}{Et} \begin{bmatrix} 1 & -v & 0 \\ -v & 1 & 0 \\ 0 & 0 & 2(1+v) \end{bmatrix} \begin{bmatrix} n_{xx} \\ n_{\theta\theta} \\ n_{x\theta} \end{bmatrix}. \tag{4.10}$$

Having obtained the strains, we compute the displacements from Eq. (4.2):

$$u_x = \int \varepsilon_{xx} dx$$

$$u_\theta = \int \left(\gamma_{x\theta} - \frac{1}{a} \frac{\partial u_x}{\partial\theta} \right) dx \tag{4.11}$$

$$u_z = a \left(\varepsilon_{\theta\theta} - \frac{1}{a} \frac{\partial u_\theta}{\partial\theta} \right)$$

These integrations also yield two auxiliary constants, but as will be shown in Sect. 4.6, these cannot always be included correctly. This is a result of the assumptions and approximations confined to the membrane theory, and thus it exhibits a limitation of this theory.

4.6 Applications of Circular Cylindrical Shells as Beam

In some simple and important cases, it is possible to introduce the boundary conditions into Eq. (4.9) and to determine the constants before making any decision regarding the particular kind of loading. All these cases have in common that $p_x = 0$ and that the other two load components, p_θ and p_z, are independent of the ordinate x. The load p_x is set equal to zero because the response can often be computed separately from the other responses. Only load cases that lead to the deflection of the circular cylindrical shell as a whole are taken into consideration. This is done because of the shortcomings of the membrane theory. If, for example, a certain type of load produces a significant ovalisation of the cross-sectional profile, the membrane theory cannot be valid and a bending theory must be used. For the described load, Eq. (4.9) yields the expressions:

$$n_{\theta\theta} = ap_z$$

$$n_{x\theta} = -\left(p_\theta + \frac{dp_z}{d\theta}\right)x + f_1(\theta)$$

$$n_{xx} = \frac{1}{2a}\left(\frac{dp_\theta}{d\theta} + \frac{d^2p_z}{d\theta^2}\right)x^2 - \frac{1}{a}\frac{df_1(\theta)}{d\theta}x + f_2(\theta)$$

$$(4.12)$$

As mentioned in Sect. 4.5, we will deal only with boundary conditions along an edge $x = constant$; note that the unknown functions are independent of the co-ordinate x. If the load components are described more explicitly, we can determine the constants. For a full circular cylinder, the load must be described by a symmetric function of the ordinate θ. Here we investigate load components described by a single sine or cosine in circumferential direction. If the load is a symmetric function, the stress resultants must also be symmetric and therefore, according to the expressions (4.12), the load components p_θ and p_z cannot be described by the same trigonometric function. The load components are thus described by the amplitudes $\hat{p}_\theta$ and $\hat{p}_z$ times the corresponding trigonometric function. By choosing the load direction in the same plane as $\theta = 0$, they become

$$p_x = 0,$$
$$p_\theta = \hat{p}_\theta \sin\theta,$$
$$p_z = \hat{p}_z \cos\theta$$

$$(4.13)$$

Substituting this description of the load components into Eq. (4.12) yields the general solution for the membrane forces in a closed circular cylindrical shell. By rewriting the general solution, we find two constants of integration, and the solution is:

$$n_{\theta\theta} = a\hat{p}_z \cos\theta$$
$$n_{x\theta} = [(\hat{p}_z - \hat{p}_\theta)x + C_1]\sin\theta$$
$$n_{xx} = -\frac{1}{a}\left[\frac{1}{2}(\hat{p}_z - \hat{p}_\theta)x^2 + C_1x + C_2\right]\cos\theta$$

$$(4.14)$$

Note that the functions $f_1(\theta)$ and $f_2(\theta)$ in Eq. (4.12) have been chosen consistently with Eq. (4.13):

$$f_1(\theta) = C_1 \sin \theta$$
$$f_2(\theta) = C_2 \cos \theta \tag{4.15}$$

The displacement components can be expressed in terms of these membrane forces by substituting the constitutive relation (4.10) into the expressions for the displacement components (4.11); this yields

$$Et\,u_x = \int n_{xx}dx - v\int n_{\theta\theta}dx$$

$$Et\,u_\theta = -\frac{1}{a}\int\int\frac{\partial n_{xx}}{\partial\theta}dxdx + \frac{v}{a}\int\int\frac{\partial n_{\theta\theta}}{\partial\theta}dxdx + 2(1+v)\int n_{x\theta}dx$$

$$Et\,u_z = \frac{1}{a}\int\int\frac{\partial^2 n_{xx}}{\partial\theta^2}dxdx - \frac{v}{a}\int\int\frac{\partial^2 n_{\theta\theta}}{\partial\theta^2}dxdx - \underline{2(1+v)\int\frac{\partial n_{x\theta}}{\partial\theta}dx} - \underline{va\,n_{xx}} + \underline{an_{\theta\theta}}$$

$$\tag{4.16}$$

On the basis of the general solution in Eq. (4.14) for the membrane forces it can be shown that the underlined terms are of the same order with respect to the axial co-ordinate x. Substitution of the general solution in these expressions for the displacements finally leads to

$$Et\,u_x = -\frac{1}{a}\left[\frac{1}{6}(\hat{p}_z - \hat{p}_\theta)x^3 + \frac{1}{2}C_1 x^2 + C_2 x + C_3\right]\cos\theta$$
$$\qquad - va[\hat{p}_z x + C_1]\cos\theta$$

$$Et\,u_\theta = -\frac{1}{a^2}\left[\frac{1}{24}(\hat{p}_z - \hat{p}_\theta)x^4 + \frac{1}{6}C_1 x^3 + \frac{1}{2}C_2 x^2 + C_3 x + C_4\right]\sin\theta$$
$$\qquad + (2+v)\left[\frac{1}{2}\left(\hat{p}_z - \frac{2(1+v)}{(2+v)}\hat{p}_\theta\right)x^2 + C_1 x + C_2\right]\sin\theta \tag{4.17}$$

$$Et\,u_z = \frac{1}{a^2}\left[\frac{1}{24}(\hat{p}_z - \hat{p}_\theta)x^4 + \frac{1}{6}C_1 x^3 + \frac{1}{2}C_2 x^2 + C_3 x + C_4\right]\cos\theta$$
$$\qquad - 2\left[\frac{1}{2}\left(\hat{p}_z - \frac{(2+v)}{2}\hat{p}_\theta\right)x^2 + C_1 x + C_2\right]\cos\theta + a^2\hat{p}_z\cos\theta$$

These expressions appear to be rather obscure. If Poisson's ratio is chosen equal to zero ($v = 0$), these expressions allow more insight.

$$Et\,u_x = -\frac{1}{a}\left[\frac{1}{6}(\hat{p}_z - \hat{p}_\theta)x^3 + \frac{1}{2}C_1 x^2 + C_2 x + C_3\right]\cos\theta$$

$$Et\,u_\theta = -\frac{1}{a^2}\left[\frac{1}{24}(\hat{p}_z - \hat{p}_\theta)x^4 + \frac{1}{6}C_1 x^3 + \frac{1}{2}C_2 x^2 + C_3 x + C_4\right]\sin\theta$$
$$\qquad + 2\left[\frac{1}{2}(\hat{p}_z - \hat{p}_\theta)x^2 + C_1 x + C_2\right]\sin\theta$$

$$Et\,u_z = \frac{1}{a^2}\left[\frac{1}{24}(\hat{p}_z - \hat{p}_\theta)x^4 + \frac{1}{6}C_1x^3 + \frac{1}{2}C_2x^2 + C_3x + C_4\right]\cos\theta$$

$$- 2\left[\frac{1}{2}(\hat{p}_z - \hat{p}_\theta)x^2 + C_1x + C_2\right]\cos\theta + a^2\hat{p}_z\cos\theta \qquad (4.18)$$

Hereafter we will discuss examples on the basis of these equations.

4.6.1 Simply Supported Tube Beam with Diaphragms

Consider the circular cylinder of span l in Fig. 4.3. This tube-shaped girder is supported by diaphragms at both ends. We consider the diaphragm infinitely rigid in its plane and perfectly flexible for deformation out-of-plane. The in-plane rigidity implies that the displacements u_θ and u_z are zero at the shell ends, and the extreme out-of-plane flexibility implies that the membrane force n_{xx} is zero at the ends of the shell. We count the co-ordinate x from the profile halfway between its ends. Because of symmetry considerations, and because of the linearity of $n_{x\theta}$ with respect to the co-ordinate x, $n_{x\theta}$ must be equal to zero halfway along the cylinder. Because of the parabolic distribution of n_{xx} and the incapability of transmitting forces in the x-direction of the diaphragms, the lengthwise distribution of n_{xx} is also fixed as shown in Fig. 4.3. The boundary conditions for the stress resultants in this shell on two diaphragms are thus described by:

$$\begin{aligned} n_{x\theta} &= 0 & at & \quad x = 0 \\ n_{xx} &= 0 & at & \quad x = \pm l/2 \end{aligned} \qquad (4.19)$$

The use of the membrane theory can be exemplified by introducing the own weight of the cylinder. This uniform load p is statically equivalent to surface load

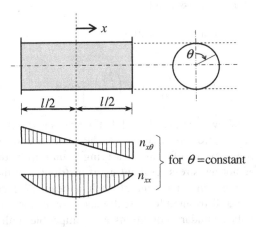

Fig. 4.3 *Circular cylindrical* shell supported by two end diaphragms

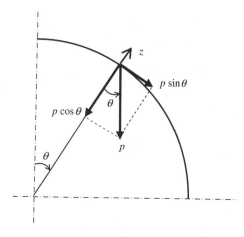

Fig. 4.4 Surface loads due to own weight

components p_θ and p_z, which are lengthwise uniformly distributed, as indicated in Fig. 4.4. These load components are thus expressed as:

$$p_\theta = p\sin\theta \Rightarrow \quad \hat{p}_\theta = p$$
$$p_z = -p\cos\theta \Rightarrow \quad \hat{p}_z = -p \tag{4.20}$$

We can evaluate the expressions in Eq. (4.14) for the membrane forces after introduction of the boundary conditions. At the ends of the girder the normal membrane force $n_{xx} = 0$, and at mid span the shear membrane force $n_{x\theta} = 0$. From this we find the constants of integration

$$C_1 = 0; \quad C_2 = -\frac{1}{8}l^2(\hat{p}_z - \hat{p}_\theta) = \frac{1}{4}pl^2 \tag{4.21}$$

The expressions for the stress resultants are

$$n_{\theta\theta} = -ap\cos\theta$$
$$n_{x\theta} = -2px\sin\theta$$
$$n_{xx} = -\frac{pl^2}{4a}\left[1 - \left(\frac{2x}{l}\right)^2\right]\cos\theta \tag{4.22}$$

These expressions show (see also Fig. 4.3) that the lengthwise distribution of the longitudinal shearing stress resultant $n_{x\theta}$ is the same as that of the transverse shear force of a simple beam of span l carrying a uniformly distributed load. Correspondingly, the normal stress resultant n_{xx} is distributed in the x-direction in the same way as the bending moment of such a beam. We conclude that a cylindrical shell under a distributed load, really acts like a simple beam loaded by a transverse load if the boundary conditions are compatible with the membrane

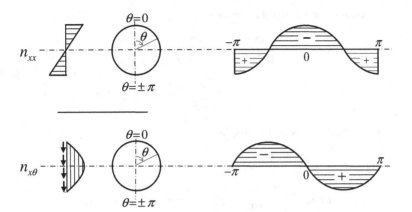

Fig. 4.5 Distribution of the membrane forces over the cross-section

assumptions, and if the profile of the shell is not distorted too much. Two interesting values of the membrane forces are:

$$n_{x\theta} = -pl\sin\theta \qquad at\ x = l/2$$
$$n_{xx} = -\frac{pl^2}{4a}\cos\theta \qquad at\ x = 0 \tag{4.23}$$

Figure 4.5 shows the distribution of the membrane forces of Eq. (4.23) over the cross-section.

The reader who will check these membrane forces, may consider the shell as a tube-shaped simply-supported girder, apply the classical Euler–Bernoulli beam theory, and will obtain the same solution. The normal stresses σ_{xx} vary linearly over the cross-section height, and the shear stresses $\sigma_{x\theta}$ parabolically.

Now we can calculate the displacements. The boundary conditions in this shell are $u_x = 0$ at $x = 0$ as this is a plane of symmetry, and $u_\theta = 0$ at $x = l/2$ due to the diaphragm. We use Eq. (4.18) for the displacements with $v = 0$ and account for these boundary conditions. Then the constants C_3 and C_4 can be calculated:

$$C_3 = 0; \quad C_4 = \frac{5}{384}l^4(\hat{p}_z - \hat{p}_\theta) = -\frac{5}{192}pl^4. \tag{4.24}$$

The expressions for the displacements u_x and u_θ are:

$$Et\,u_x = -\frac{pl^2}{24a}\left(\frac{2x}{l}\right)\left[3 - \left(\frac{2x}{l}\right)^2\right]\cos\theta$$

$$Et\,u_\theta = \frac{pl^4}{192a^2}\left[5 - 6\left(\frac{2x}{l}\right)^2 + \left(\frac{2x}{l}\right)^4\right]\sin\theta \tag{4.25}$$

$$+ \frac{1}{2}pl^2\left[1 - \left(\frac{2x}{l}\right)^2\right]\sin\theta$$

The expression for the displacement u_z is:

$$Et\,u_z = -\frac{pl^4}{192a^2}\left[5 - 6\left(\frac{2x}{l}\right)^2 + \left(\frac{2x}{l}\right)^4\right]\cos\theta$$

$$-\frac{1}{2}pl^2\left[1 - \left(\frac{2x}{l}\right)^2\right]\cos\theta - a^2 p\cos\theta$$

(4.26)

Two interesting values of the displacements are

$$Et\,u_x = -\frac{pl^2}{12a}\cos\theta \qquad at\ x = l/2$$

$$Et\,u_\theta = \frac{pl^4}{a^2}\left[\frac{5}{192} + \frac{1}{2}\left(\frac{a}{l}\right)^2\right]\sin\theta \qquad at\ x = 0$$

(4.27)

The value of the displacement u_z at the same profiles is

$$Et\,u_z = -\frac{pl^4}{a^2}\left\{\left[\frac{5}{192} + \frac{1}{2}\left(\frac{a}{l}\right)^2\right] + \left(\frac{a}{l}\right)^4\right\}\cos\theta \qquad at\ x = 0$$

$$Et\,u_z = -a^2 p\cos\theta \qquad at\ x = l/2$$

(4.28)

The displacement u_θ in a cross-section is maximal at $\theta = \pi/2$ (at the horizontal line through the circle centre), and the displacement u_z at $\theta = 0$ (at the vertical line through the circle centre). It is obvious that for a slender beam with a length much larger than the depth ($l > 2a$) the difference between the displacement u_θ and u_z in the middle of the span ($x = 0$) is of the order $(a/l)^4$ and therefore negligible. So they are equal for a slender beam of circular cross-section. This means that no ovalisation occurs for this load. The reader who will check the deflection w of the beam by the classical bending Euler-Bernoulli theory for beams, will obtain $Et\,w = 5pl^4/(192a^2)$ at the middle of the span. In Eqs. (4.27) and (4.28) the factor $5/192$ has become $5/192 + (a/l)^2/2$. The difference is due to shear deformation. The contribution to the deflection due to shear deformation is related to the bending deformation as $(a/l)^2$ to unity. For slender beams a/l is small, and the shear contribution is negligible.

The displacement u_z normal to the surface is uniformly distributed over the cross-section at the ends of the beam. This is in conflict with the assumption that the diaphragms at the supports are perfectly rigid for in-plane deformation. In the next chapter we will learn that additional bending stresses occur in an edge zone, and that these are negligible compared to the stresses due to the membrane action of the shell.

The influence of the lateral contraction v is also negligible for the beam of this example provided that $l > 2a$. Without going into details we find that for a non-zero Poisson's ratio the constants derived by the boundary conditions for the displacements are:

$$C_3 = 0; \quad C_4 = \left(-\frac{5}{192} + v\frac{1}{8}\left(\frac{a}{l}\right)^2\right)pl^4$$

(4.29)

On basis of this solution we conclude that the influence of the lateral contraction is approximately of the order $(a/l)^2$ compared to unity, which is negligible for slender beams.

4.6.2 Circular Shell as a Cantilever Beam

We consider a cantilever beam with a circular cross-section and analyze it as a shell. The shell in Fig. 4.6 is completely built in at $x = l$. The support is able to resist not only the shear membrane force $n_{x\theta}$ but also the normal membrane force n_{xx} and therefore represents another case of boundary conditions than in Sect. 4.6.1. At this clamped edge the displacements u_x and u_θ are zero, and at the free edge there are no membrane forces. Hence, the boundary conditions are:

$$\begin{aligned} n_{x\theta} &= 0 \quad at \quad x = 0 \\ n_{xx} &= 0 \quad at \quad x = 0 \\ u_x &= 0 \quad at \quad x = l \\ u_\theta &= 0 \quad at \quad x = l \end{aligned} \tag{4.30}$$

By evaluating Eq. (4.14) for the membrane forces after introduction of the first two boundary conditions, we find that the constants of integration C_1 and C_2 are zero and hence the expressions for the stress resultants are

$$\begin{aligned} n_{\theta\theta} &= a\hat{p}_z \cos\theta \\ n_{x\theta} &= (\hat{p}_z - \hat{p}_\theta)x \sin\theta \\ n_{xx} &= -\frac{1}{2a}(\hat{p}_z - \hat{p}_\theta)x^2 \cos\theta \end{aligned} \tag{4.31}$$

Supported as a cantilever beam, the lengthwise distributions of z and σ_{zz} in this shell are those of the shear force and the bending moment of the beam analogy.

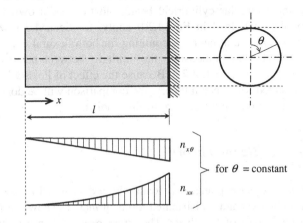

Fig. 4.6 Circular cylindrical shell as cantilever beam

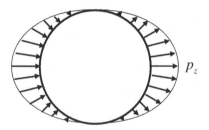

Fig. 4.7 Simple representation of wind load. Pressure *left*, suction *right*

In this example we choose a load such that u_z and p_θ are equal to zero. The load p_z is a uniformly distributed load p_i in the A-direction and is distributed in circumferential direction as a cosine function:

$$p_z = -p \cos\theta \Rightarrow \hat{p}_z = -p \tag{4.32}$$

For a vertical cylinder this load can be considered as a simple representation of a wind load. The pressure on the front side and the suction on the back side are equal, see Fig. 4.7.

In Sect. 10.1 we will work with a more sophisticated representation of the wind pressure distribution around a cylinder. Substitution of Eq. (4.32) in Eq. (4.31) yields the solution

$$n_{x\theta} = -px \sin\theta; \quad nxx = \frac{p}{2a}x^2 \cos\theta. \tag{4.33}$$

4.7 Circular Beam Under a Transverse Load

As shown in Sect. 4.6 an analogy exists between the presented membrane solution and the behaviour of a transversely loaded beam. The objective of this paragraph is to show the differences and the similarities between the membrane solution and the beam solution for a circular cylindrical beam and the load's own weight. The expressions for the shell load in Eq. (4.20) apply. In Sect. 4.7.1 we will first discuss the problem in beam theory, accounting for both flexural deformation and shear deformation. Then we demonstrate the similarity between beam solution and shell membrane solution in Sect. 4.7.2. Because the effect of Poisson's ratio is not consistently accounted for in beam theory, the comparison will be done for $v = 0$. At the end of the section we will return to this choice.

4.7.1 Solution in Beam Theory

Consider a beam in the x–z plane. It is loaded by a distributed load $q(x)$ perpendicular to its middle axis and a distributed moment $m(x)$ as shown in Fig. 4.8 for an elementary beam part of length dx. The stress resultants in a section are the

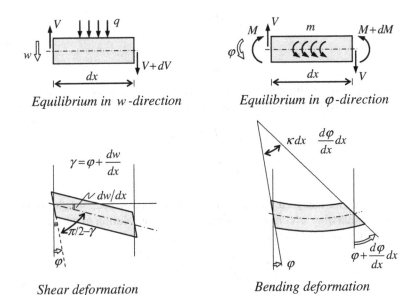

Fig. 4.8 Definitions for beam element with flexural and shear deformation

bending moment M and the transverse shear force V. The lateral displacement $n_{ij} = \int_t \sigma_{ij}(z)dz$ is the displacement of the middle axis of the cross-section in the z-direction and φ is the rotation of the cross-section. The equilibrium equations for the infinitesimal element are

$$\frac{dV}{dx} + q = 0$$

$$\frac{dM}{dx} - V + m = 0 \tag{4.34}$$

For a beam having resistance to bending as well as shearing, the moment is related to the curvature κ and the transverse shear is related to the shear angle δe^T. The elastic constants are respectively the flexural rigidity EI and the shear rigidity GA_s. For a thin circular beam the moment of inertia $\mathbf{u}$ is equal to half of the polar moment of inertia and the effective shear cross-section A_s is related to the cross-section $\mathbf{B}^T\mathbf{s} = \mathbf{p}$ by $A_s = \eta A$. For a ring-shaped cross-section it holds that $\eta = 2$. The constitutive equations are thus described by:

$$M = EI\kappa$$

$$V = GA_s \tag{4.35}$$

The shear angle γ is described by the turning over dw/dx of the middle line of the beam due to the displacement w and the rotation $\mathbf{B}$ of the cross-section. The kinematic equations are thus described by:

$$\gamma = \frac{dw}{dx} + \varphi$$

$$\kappa = \frac{d\varphi}{dx} \tag{4.36}$$

Substituting Eq. (4.36) into Eq. (4.35) relates the transverse shear and the bending moment to the degrees of freedom by

$$V = GA_s \left(\frac{dw}{dx} + \varphi \right)$$

$$M = EI \frac{d\varphi}{dx} \tag{4.37}$$

For the cases we analysed with the membrane theory of Sect. 4.6, it holds that $m(x) = 0$ and that $q(x) = constant$. For these cases, the solution to the equilibrium equations of Eq. (4.34) becomes

$$V = \frac{dM}{dx} = -\int q dx$$

$$M = -\iint q dx dx \tag{4.38}$$

Naturally the lengthwise distribution of the transverse shear is linear and the distribution of the bending moment is parabolic with respect to the axial co-ordinate x. By comparing the membrane solution of Eq. (4.31) with this solution, the beam analogy is already distinguished.

The solution for the rotation is found by substituting solution Eq. (4.38) into Eq. (4.37) for the bending moment. It is, for a constant EI,

$$\varphi = -\frac{1}{EI} \iiint q dx dx dx \tag{4.39}$$

The solution for the displacement is found as follows. We eliminate V from the first equation of (4.37) and the first equation of (4.38). The newly found equation is integrated with respect to x:

$$w = -\frac{1}{EI} \int \varphi \, dx - \frac{1}{GA_s} \iint q \, dx dy. \tag{4.40}$$

The final result is

$$V = \frac{dM}{dx} = -\int q dx = -[qx + c_1]$$

$$M = -\iint q dx dx = -[\frac{1}{2} qx^2 + c_1 x + c_2]$$

$$\varphi = -\frac{1}{EI} [\frac{1}{6} qx^3 + \frac{1}{2} c_1 x^2 + c_2 x + c_3] \tag{4.41}$$

$$w = \frac{1}{EI} [\frac{1}{24} qx^4 + \frac{1}{6} c_1 x^3 + \frac{1}{2} c_2 x^2 + c_3 x + c_4] - \frac{1}{GA_s} [\frac{1}{2} qx^2 + c_1 x + c_2]$$

4.7.2 Comparison of Beam Solution with Shell Membrane Solution

The result in Eq. (4.41) is compared with the general solution for the membrane forces and the displacements in the membrane theory of a circular cylindrical shell as shown in Eqs. (4.14) and (4.18). Noticing the distribution as function of x, it appears that the quantities of the beam theory are related to the quantities of the membrane theory as follows:

$$
\begin{aligned}
V &\leftrightarrow n_{x\theta} & \varphi &\leftrightarrow u_x \\
M &\leftrightarrow n_{xx} & w &\leftrightarrow u_\theta
\end{aligned}
\tag{4.42}
$$

Firstly we will derive the stresses in a thin circular cross-section due to n_{yx} and v_{xz} and, secondly, we will express the beam displacements v_{yz} and w in terms of displacements at the middle surface of the circular cylinder.

Moment and Shear Force

To calculate stress distribution due to surface loads by means of the theory of beams, the load distribution around the circumference must be integrated. The shell loads and the beam load are shown in their co-ordinate systems in Fig. 4.9. The expression for the equivalent transverse load m_{yy} becomes

$$
q = 2 \int_0^\pi (p_\theta \sin\theta - p_z \cos\theta) a\,d\theta
\tag{4.43}
$$

Because $p_\theta = \hat{p}_\theta \sin\theta$ and $p_z = \hat{p}_z \cos\theta$ we obtain

$$
q = 2 \int_0^\pi \left(\hat{p}_\theta \sin^2\theta - \hat{p}_z \cos^2\theta\right) a\,d\theta
\tag{4.44}
$$

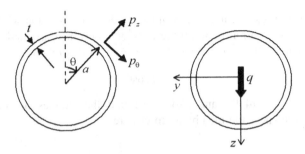

Fig. 4.9 Shell loads in polar and beam load in Cartesian co-ordinate system

Substitution of the well-known equality $\int\limits_0^\pi \sin^2\theta d\theta = \int\limits_0^\pi \cos^2\theta d\theta = \frac{1}{2}\pi$ yields

$$q = (\hat{p}_\theta - \hat{p}_z)a\pi \tag{4.45}$$

Substitution of the load expressions of Eq. (4.13) yields a value $q = 2\pi a p$ for its own weight per unit length of the cylinder. We could have made a short cut by multiplying the circumference of the cylinder $2\pi a$ by its own weight p per unit area of the cylinder.

The stresses due to the transverse shear and the bending moment can be calculated with the aid of the simple beam formulas:

$$\begin{aligned}
\sigma_{x\theta} &= -\frac{V \cdot S_z^{(s)}}{b^{(s)} \cdot I} \\
\sigma_{xx} &= \frac{M \cdot z}{I}
\end{aligned} \tag{4.46}$$

In these formulas I is the moment of inertia, $S_z^{(s)}$ and $b^{(s)}$ are the first moment of area and the width of the shearing section respectively and z is the vertical co-ordinate. For a ring, the moment of inertia is half the polar moment of inertia and the width of the shearing section is twice the thickness. Another important property of the cross-section is the cross-sectional area A. The cross-sectional quantities are thus

$$b^{(s)} = 2t$$

$$A = \int\limits_0^{2\pi} ta\,d\theta = 2\pi at$$

$$S_z^{(s)} = -2\int\limits_0^\theta a\cos\theta \cdot ta\,d\theta = -2a^2t\sin\theta \tag{4.47}$$

$$I = \frac{1}{2}I_{polar} = \frac{1}{2}\int\limits_0^{2\pi} a^2 \cdot ta\,d\theta = \pi a^3 t$$

To obtain useful expressions for comparing the beam solution with the membrane solution, the relation between the angular co-ordinate ds_y and the vertical beam co-ordinate ds_y is set up. According to Fig. 4.9 this relation is

$$z = -a\cos\theta \tag{4.48}$$

From Eqs. (4.46), (4.47) and (4.48), we find the stresses in beam theory. Multiplied by thickness the membrane forces are

$$n_{x\theta} = \sigma_{x\theta}t = \frac{V}{\pi a}\sin\theta$$

$$n_{xx} = \sigma_{xx}t = \frac{M}{\pi a^2}\cos\theta \tag{4.49}$$

This solution of beam theory differs with membrane theory only in a multiplication factor. The distribution in x-direction is the same. Therefore, the expressions found from the beam theory are identical to the general solution to the membrane theory.

Displacements

From Fig. 4.10 we see that the beam displacements φ and w can be expressed in terms of shell displacements at the middle surface: u_x, n_{xy} and v_{xz}. Using Eq. (4.48) for z, we find the expressions

$$\begin{array}{ll} \varphi \to u_x & u_x = \varphi \cdot z = -a\varphi\cos\theta \\[2mm] w \to \begin{array}{l} u_\theta \\ u_z \end{array} & \begin{array}{l} u_\theta = w\sin\theta \\ u_z = -w\cos\theta \end{array} \end{array} \tag{4.50}$$

We substitute the general solution Eq. (4.41) into these expressions and find

$$EI\,u_x = a\left[\frac{1}{6}qx^3 + \frac{1}{2}c_1x^2 + c_2x + c_3\right]\cos\theta$$

$$EI\,u_\theta = \left[\frac{1}{24}qx^4 + \frac{1}{6}c_1x^3 + \frac{1}{2}c_2x^2 + c_3x + c_4\right]\sin\theta$$

$$\qquad - \frac{EI}{GA_d}\left[\frac{1}{2}qx^2 + c_1x + c_2\right]\sin\theta \tag{4.51}$$

$$EI\,u_z = -\left[\frac{1}{24}qx^4 + \frac{1}{6}c_1x^3 + \frac{1}{2}c_2x^2 + c_3x + c_4\right]\cos\theta$$

$$\qquad + \frac{EI}{GA_d}\left[\frac{1}{2}qx^2 + c_1x + c_2\right]\cos\theta$$

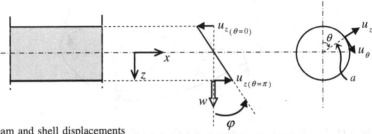

Fig. 4.10 Beam and shell displacements

The constants c_1, c_2. c_3 and c_4 must follow from the support conditions at both ends of the cylinder. Noticing that $I = \pi a^3 t$ and $q = (\hat{p}_\theta - \hat{p}_z)a\pi$, this beam theory result is completely equal to the expressions which we have obtained in Eq. (4.18) on the basis of shell membrane theory. In comparison with membrane theory, beam theory is not able to predict the magnitude and distribution of the stress resultant σ_{xy} in the circumferential direction and does not take into account the cross-sectional deformation due to the lateral contraction z. As shown in Sect. 4.6.1 these differences are of no importance if the circular beam is slender $(l > 2a)$. For such a beam the membrane forces m_{yx} and $(r_x + z)/r_x$ are the important stress resultants. For stocky cylinders, the membrane theory has to be used to find appropriate solutions for the stresses and the cross-sectional deformation.

4.8 In-Extensional Deformation of a Circular Storage Tank

Storage tanks will be the extensive subject of Chap. 13, where both membrane and bending forces will be included. Here we restrict the discussion to a special case of pure membrane action. We consider a circular cylindrical tank with a base plate and a floating-cover-system. The base plate can be considered inextensible in comparison with the deformable shell. For such tanks, a non-uniform settlement of the foundation often results in jamming of the cover. It turns out that at some height of the cylinder the shape of the cross-section becomes elliptic, see Fig. 4.11, which is at first sight an unlikely consequence of the vertical displacements of the base. It can happen because of in-extensional deformations: the shell is deformed while the strains of the middle surface remain zero. A shell, as a thin-walled

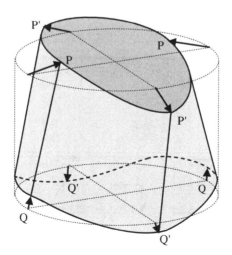

Fig. 4.11 In-extensional deformation of a circular cylindrical shell

structure, which is much stiffer in-plane than perpendicular to its plane, has a strong preference for such deformations. The in-extensional deformation of the cylinder is described by a relatively simple formulation. The strains of the middle surface are expressed in terms of the displacements and the kinematical relation Eq. (4.2):

$$\varepsilon_{xx} = \frac{\partial u_x}{\partial x}$$
$$\varepsilon_{\theta\theta} = \frac{1}{a}\left(\frac{\partial u_\theta}{\partial \theta} + u_z\right) \qquad (4.52)$$
$$\gamma_{x\theta} = \frac{1}{a}\frac{\partial u_x}{\partial \theta} + \frac{\partial u_\theta}{\partial x}$$

Observing the deformed shape of the shell in Fig. 4.11, we see that the base plane and the top plane of the shell are warped. Displacements in the vertical direction seem independent of the coordinate x, but highly dependent on θ. Over the full circumference the vertical displacement is two times positive and two times negative. This indicates two waves, so $n = 2$. For the displacement u_x, we choose the shape $u_x(\theta) = C_x a \cos 2\theta$, where C_x is a dimensionless constant. So, the vertical displacement is independent of the x-coordinate. Figure 4.11 shows that the normal displacement u_z at the top plane of the shell becomes two times positive and two times negative, with maxima in the points P and P'. The function $\cos 2\theta$ applies again for the circumferential direction, however, this displacement is not constant over the height; it varies from zero at the base (due to the inextensible base plate) to a nonzero value at the top. We choose a linear variation over the height, so the function is $u_z(x, \theta) = C_z x \cos 2\theta$. Similar arguments hold for the circumferential displacement u_θ, with the difference that it is zero at P and P' and maximal at Q and Q'. Therefore we change cosine to sine: $u_\theta(x, \theta) = C_\theta x \sin 2\theta$. If the three constants are C_x, C_θ and C_z are chosen in proportion 1 to 2 to -4, we arrive at a special set:

$$u_x(x, \theta) = Ca \cos(2\theta)$$
$$u_\theta(x, \theta) = 2Cx \sin(2\theta) \qquad (4.53)$$
$$u_z(x, \theta) = -4Cx \cos(2\theta)$$

For this combination of displacements all three strains in Eq. (4.52) become zero:

$$\varepsilon_{xx} = 0, \quad \varepsilon_{\theta\theta} = 0, \quad \gamma_{x\theta} = 0. \qquad (4.54)$$

The displacement field of Eq. (4.53) fully describes the in-extensional phenomenon which is shown in Fig. 4.11. Note that the normal displacement u_z at the top is as large as four times the vertical settlement of the base A_0. A limited unevenness at the base leads to a substantial unroundness at the top, which easily hinders the floating roof cover to move vertically.

4.9 Circular Shell Under an Axisymmetric Load

A circular cylindrical shell is often used as a tank for storing liquid or gas. In these cases the shell is subject to an axisymmetric load, meaning that the normal load component p_z, for $x = constant$, is uniform and that the circumferential load component A is zero. Since the intensity of the normal load component depends only on the ordinate x, the normal displacement φ_θ is also uniform along the circumference. Furthermore, all the derivatives with respect to the circumferential co-ordinate θ are zero. The circumferential displacement u_θ and the longitudinal shearing membrane force $n_{x\theta}$ are also zero, because of symmetry considerations. Setting the load component A equal to zero for convenience and inspecting the rewritten equilibrium equations (4.7) shows that the normal membrane force A_0 is also zero.

For an axisymmetric load $p_z(x, \theta) = p_z(x)$ the membrane forces are

$$
\begin{aligned}
n_{\theta\theta} &= p_z(x)a \\
n_{x\theta} &= 0 \\
n_{xx} &= 0
\end{aligned}
\qquad (4.55)
$$

Substitution of this result in the constitutive relations of Eq. (4.10) yields

$$
\begin{aligned}
\varepsilon_{xx} &= \frac{1}{Et}(n_{xx} - \nu n_{\theta\theta}) = -\nu\frac{p_z(x)a}{Et} \\
\varepsilon_{\theta\theta} &= \frac{1}{Et}(-\nu n_{xx} + n_{\theta\theta}) = \frac{p_z(x)a}{Et} \\
\gamma_{x\theta} &= \frac{2(1+\nu)}{Et}n_{x\theta} = 0
\end{aligned}
\qquad (4.56)
$$

Successive substitution of this result in the rewritten kinematic relations of Eq. (4.11) yields

$$
\begin{aligned}
u_x &= \int \varepsilon_{xx}dx & &= -\nu\frac{a}{Et}\int p_z(x)dx + A \\
u_\theta &= \int \left(\gamma_{x\theta} - \frac{1}{a}\frac{\partial u_x}{\partial\theta}\right)dx & &= B \\
u_z &= a\left(\varepsilon_{\theta\theta} - \frac{1}{a}\frac{\partial u_\theta}{\partial\theta}\right) & &= \frac{p_z(x)a^2}{Et}
\end{aligned}
\qquad (4.57)
$$

The constant of integration A can be evaluated with the aid of the boundary conditions, and the second constant of integration B must be zero, because u_θ is zero for axisymmetric loads. This membrane analysis of the axisymmetric case is applicable only if the displacement u_x is fixed at one end of the shell, and if the increase of the radius to $a + u_z$ is not prevented.

4.9.1 Application to Water Tank

The circular cylindrical tank of Fig. 4.12 is completely filled with water with density γ, and is supported by roller supports. Since the upper edge is free and the lower edge does not withstand an increase of the radius as well, displacements can freely develop and do not impose boundary conditions. Therefore, the membrane theory is applicable. The shell is subjected to a normal load:

$$p_z = \gamma(l - x) \tag{4.58}$$

The stress resultant $n_{\theta\theta}$ in the circumferential direction is the only membrane response to this axisymmetric load and is determined by Eq. (4.14):

$$n_{\theta\theta} = \gamma a(l - x) \tag{4.59}$$

The resulting strains are computed by Eq. (4.10), yielding

$$\varepsilon_{\theta\theta}(xx) = -v\frac{\gamma a}{Et}(l - x); \quad \varepsilon_{\theta\theta}(xx) = \frac{\gamma a}{Et}(l - x) \tag{4.60}$$

Before deriving the displacement with the aid of Eq. (4.11), we set up the boundary condition for the displacement at the edge $x = 0$. The displacement u_x should be zero along this edge since a vertical displacement is prohibited here. When this boundary condition is taken into account, all integration constants become zero, and the expressions for the longitudinal and the normal displacement are

$$u_x = -v\frac{\gamma a}{Et}\left(lx - \frac{1}{2}x^2\right)$$
$$u_z = \frac{\gamma a^2}{Et}(l - x) \tag{4.61}$$

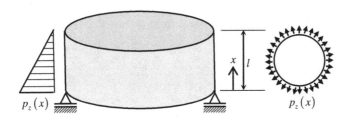

Fig. 4.12 Circular cylindrical tank loaded by water pressure

4.10 Concluding Remarks

Membrane theory is a straightforward and useful tool for analysing the stress distribution in circular cylindrical shells under surface loading. Since the stress resultants can be computed directly by the equilibrium equations, it is not necessary to take into account the deformation of the structure.

In this chapter, we have shown that the membrane theory is applicable to closed circular cylindrical shells whose cross-sectional profiles are closed curves and which have only two cross-sectional profiles as boundaries. Not all types of surface loading can be considered since membrane theory cannot accommodate considerable bending of the shell surface in either the axial or the tangential direction.

Two load types of closed circular cylinders were analysed:

1. The lateral load expressed by $p_\theta(x, \theta) = \hat{p}_\theta \sin\theta$, $p_z(x, \theta) = \hat{p}_z \cos\theta$;
2. The axisymmetric load expressed by $p_z(x, \theta) = \hat{p}_z$.

The distribution of the stress under lateral loading is analogous to the load carrying mechanism of a simple beam. Of course this is true only if it is not necessary to take into account the cross-sectional deformation, because then a bending theory must be applied. The advantage of membrane theory over beam theory is that an analysis with the membrane theory directly produces the stress distribution over the cross-sectional area. Furthermore, beam theory is not able to predict the magnitude and distribution of the circumferential stress resultant and does not take into account the influence of the lateral contraction ν.

Membrane theory is applicable only to axisymmetric loading if the boundary conditions are such that the membrane displacements are not constrained. If an edge does prevent membrane displacements, a bending analysis must be applied to solve the problem. This will be the subject of the next chapter.

Chapter 5
Edge Disturbance in Circular Cylindrical Shells Under Axisymmetric Load

5.1 Problem Assignment

As stated in Sect. 4.9, the axisymmetry of the load for a circular cylindrical shell means that the axial load component p_x and normal load component p_z are uniformly distributed along the circumference, and the circumferential load component p_θ is zero. This chapter deals with the analysis of circular cylindrical shells with edges for x = constant under axisymmetric loading. We put p_x to zero, so consider only a normal load p_z. In Sect. 4.9 we showed that the water tank must be able to perform a normal displacement due to the loading to make the membrane assumptions applicable. In reality we will have a hinged support or a clamped edge, and this prevents the normal displacement at the base of the shell. The membrane solution will be disturbed, and bending actions will have to provide the continuity of displacements. The disturbance zone will appear to be of limited size close to the edge. We show this in Fig. 5.1 for a clamped tank wall. The final result is a superposition of the membrane solution and bending solution. Apart from membrane forces, also bending moments occur. The membrane solution accounts for the distributed load p_z on the shell, and the bending solution accounts for edge loads at the base of the shell in order to satisfy the boundary constraints. Because of the simplicity of the loading and the single curvature of the cylindrical shell, the bending theory for the edge disturbance remains simple. We will derive more general shell theories, including bending, in Chaps. 6 and 9. We consider the circular cylindrical shell of Fig. 5.2, which is loaded in an axisymmetric way at edge $x = 0$ by edge effects. These are required to compensate a membrane incompatible displacement u_z. We choose the origin of the x-axis at the edge of the shell. Because of symmetry considerations, all derivatives with respect to ordinate θ are zero. For the same reason, the circumferential displacement u_θ, the shear strain $\gamma_{x\theta}$, the shear membrane force $n_{x\theta}$ and the twisting moment $m_{x\theta}$ are all zero. The normal membrane force n_{xx} can be taken as zero because the opposite edge is unconstrained.

The change of curvature κ_{xx} will not be zero, which has to vary with respect to the ordinate x in order to overcome any discrepancies between the boundary conditions at the base and the membrane displacements of the shell, see Fig. 5.1.

J. Blaauwendraad and J. H. Hoefakker, *Structural Shell Analysis*,
Solid Mechanics and Its Applications 200, DOI: 10.1007/978-94-007-6701-0_5,
© Springer Science+Business Media Dordrecht 2014

This means that the bending moment m_{xx} and the transverse shearing stress resultant v_x vary with respect to the ordinate x. Accordingly, the circumferential normal membrane force $n_{\theta\theta}$ and bending moment $m_{\theta\theta}$ are also activated; they are constant in circumferential direction because of the axisymmetry. The bending moment $m_{\theta\theta}$ can only develop due to the effect of Poisson's ratio. The curvature $\kappa_{\theta\theta}$ is zero, therefore $m_{\theta\theta} = v m_{xx}$. For zero Poisson's ratio, no moment $m_{\theta\theta}$ occurs.

5.2 Derivation of a Differential Equation

Accounting for all these expectations, we consider a circular cylindrical shell in axisymmetric bending as a barrel consisting of vertical staves and horizontal ring belts, see Fig. 5.3. The staves are straight and have unit width, and classical Euler–Bernoulli beam theory applies. The normal displacement u_z, the bending moment m_{xx} and transverse shear force v_x are functions of the coordinate x only. If the staves tend to

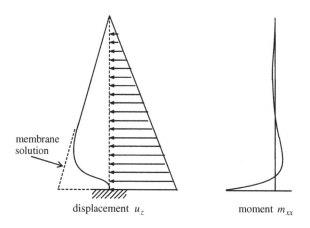

Fig. 5.1 Edge disturbance at the base of a tank

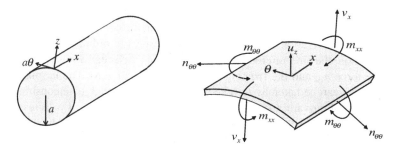

Fig. 5.2 Axisymmetric shell. Sign convention

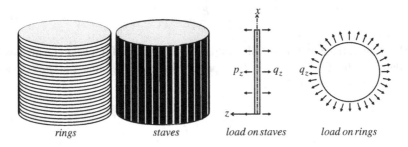

rings   *staves*   *load on staves*   *load on rings*

Fig. 5.3 Rings and staves in a circular cylindrical shell

displace outward, the rings will be strained. A uniformly distributed interaction force q_z acts outward on each ring and inward on the staves. The force q_z is proportional to the displacement u_z. The differential equation for the stave without rings is

$$D\frac{d^4 u_z}{dx^4} = p_z \tag{5.1}$$

Here D is the *flexural rigidity*

$$D = \frac{Et^3}{12(1 - v^2)} \tag{5.2}$$

The factor $1 - v^2$ is due to the fact that no change of curvature can occur in a circumferential direction. Hereafter we will put p_z to zero, because we consider only bending that occurs due to loads at the base of the shell.

We now consider the rings of unit width. Due to the normal displacement u_z the radius of the rings will increase. Load q_z acts on the rings in positive z-direction and an equal load on the staves in negative direction. Beam Eq. (5.1) changes into

$$D\frac{d^4 u_z}{dx^4} = p_z - q_z \tag{5.3}$$

We calculate load q_z as follows. The circumferential strain is $\varepsilon_{\theta\theta} = u_z/a$. Because no load p_x is present, and the top edge of the tank is free, the membrane force n_{xx} will be zero. Therefore, the rings are in a uniaxial state of stress, and the relation between the strain $\varepsilon_{\theta\theta}$ and the circumferential membrane force $n_{\theta\theta}$ is $n_{\theta\theta} = Et\,\varepsilon_{\theta\theta}$. Note that the vertical strain ε_{xx} will be $v\,\varepsilon_{\theta\theta}$, and the vertical displacement is not zero. The load on the rings is $q_z = n_{\theta\theta}/a$. Combining the three expressions for $\varepsilon_{\theta\theta}$, $n_{\theta\theta}$ and q_z, we obtain

$$\frac{Et}{a^2} u_z = q_z \tag{5.4}$$

We substitute Eq. (5.4) in Eq. (5.3) and obtain the differential equation that governs the edge disturbance.

$$D\frac{d^4u_z}{dx^4} + \frac{Et}{a^2}u_z = p_z \qquad (5.5)$$

The differential equation is fully similar to the equation for a beam on an elastic foundation. The term Et/a^2 is the spring stiffness. We now put $p_z = 0$ and introduce the parameter β:

$$\beta^4 = \frac{Et}{4Da^2} = \frac{3(1-v^2)}{(at)^2}. \qquad (5.6)$$

The homogeneous equation becomes

$$\frac{d^4u_z}{dx^4} + 4\beta^4 u_z = 0 \qquad (5.7)$$

The solution of this fourth-order differential equation is $u_z = Ce^{rx}$. Substitution in Eq. (5.7) leads to an equation for the roots r:

$$r^4 + 4\beta^4 = 0 \qquad (5.8)$$

The four roots are

$$r = \pm(1 \pm i)\beta \qquad (5.9)$$

The solution consists of two pairs of conjugate complex functions. The sum and the difference of the functions of each pair are purely real and purely imaginary and constitute another set of four independent homogenous solutions:

$$u_z(x) = e^{-\beta x}[C_1 \cos \beta x + C_2 \sin \beta x] + e^{\beta x}[C_3 \cos \beta x + C_4 \sin \beta x] \qquad (5.10)$$

In this expression, the terms with C_1 and C_2 are oscillating functions of x that decrease exponentially with increasing x. The other two terms are also attenuating oscillations but these increase exponentially with increasing x; we can also say that they decrease with decreasing x. Without loss of generality the solution of Eq. (5.10) can be written as

$$u_z(x) = e^{-\beta x}[A_1 \cos \beta x + A_2 \sin \beta x] + e^{-\beta(l-x)}[B_1 \cos \beta x + B_2 \sin \beta x] \qquad (5.11)$$

Now the first term is an attenuating disturbance, starting from the edge $x = 0$ and the second term is another one from the opposite edge $x = l$. As shown in Fig. 5.4, we in fact have introduced a new coordinate x' that starts at the edge $x = l$ and increases in the direction of the edge $x = 0$. We introduce the *characteristic length* l_c:

$$l_c = \frac{1}{\beta} = \frac{\sqrt{at}}{\sqrt[4]{3(1-v^2)}} \qquad (5.12)$$

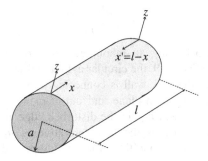

Fig. 5.4 Axial ordinates x and x'

The argument of the exponential function is x/l_c. The exponential function e^{-x/l_c} is 1 at $x = 0$ and has reduced to about 0.05 when $x/l_c = \pi$. We call this value of x the *influence length* of the *edge disturbance*. This influence length l_i is thus:

$$l_i = \pi l_c = \frac{\pi}{\sqrt[4]{3(1-v^2)}}\sqrt{at} \tag{5.13}$$

To exemplify the influence of an edge disturbance, we compare the influence length with the radius of the cylinder. This influence-length-to-radius ratio reads:

$$\frac{l_i}{a} = \frac{\pi l_c}{a} = \frac{\pi}{\sqrt[4]{3(1-v^2)}}\sqrt{\frac{t}{a}} \tag{5.14}$$

Since the thickness-to-radius ratio of a thin shell is of the order 1/100, the influence length l_i of the edge disturbance is about four times shorter than the radius of the cylinder. Storage tanks have heights l that are of the order of the diameter $2a$. So, the influence length is about one-eighth of the height. Indeed, the membrane solution is disturbed only in an edge zone at the base of the shell. If the shell has no constraints at the top, the membrane solution is not disturbed there, and the coefficients B_1 and B_2 will be zero.

5.3 Application to a Water Tank

We return to the water tank of Sect. 4.9. There the tank was placed on rollers at the base, which permitted a pure membrane displacement field and associated solution. We repeat here the equations for the load, the circumferential membrane force and the normal displacement:

$$p_z = \gamma(l - x) \tag{5.15}$$

$$n_{\theta\theta} = \gamma a(l - x) \tag{5.16}$$

$$u_z = \gamma \frac{a^2(l - x)}{Et} \tag{5.17}$$

In fact, Eqs. (5.16–5.17) are an inhomogeneous solution of the differential Eq. (5.5). In contrast to Sect. 4.9, the circular cylindrical water tank cannot deform at the lower edge since the tank wall is connected to a thick flat plate, which is assumed to be infinitely rigid in-plane and out-of-plane, see Fig. 5.5. We will determine the bending moment m_{xx} in the disturbed edge zone. For convenience, we introduce the rotation φ_x as the first derivative of u_z with negative sign: $\varphi_x = -du_z/dx$. The boundary conditions for the clamped edge at $x = 0$ are

$$u_z = 0, \quad \varphi_x = 0 \tag{5.18}$$

It means that the displacement and rotation of the membrane solution must be compensated by an opposite displacement and rotation due to an edge moment and transverse edge force. Assuming that the length of the cylinder is larger than the influence length l_i, only the first two terms of the solution in Eq. (5.10) are used. Further we find the bending moment m_{xx} from the second derivative of u_z with a minus sign, multiplied by the flexural rigidity D, and the transverse shear force v_x by differentiation of the bending moment. The result is

$$
\begin{aligned}
u_z &= e^{-\beta x}[C_1 \cos \beta x + C_2 \sin \beta x] \\
\varphi_x &= \beta e^{-\beta x}[(C_1 - C_2) \cos \beta x + (C_1 + C_2) \sin \beta x] \\
n_{\theta\theta} &= Et \frac{1}{a} e^{-\beta x}[C_1 \cos \beta x + C_2 \sin \beta x] \\
m_{xx} &= 2D \beta^2 e^{-\beta x}[C_2 \cos \beta x - C_1 \sin \beta x] \\
v_x &= -2D \beta^3 e^{-\beta x}[(C_1 + C_2) \cos \beta x - (C_1 - C_2) \sin \beta x]
\end{aligned} \tag{5.19}
$$

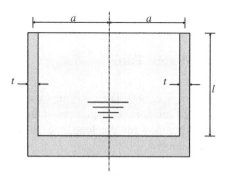

Fig. 5.5 Circular cylindrical tank connected to a thick flat plate

The complete expressions for the displacement and the rotation are

$$u_z = \gamma \frac{a^2}{Et}(l - x) + e^{-\beta x}[C_1 \cos \beta x + C_2 \sin \beta x]$$

$$\varphi_x = -\gamma \frac{a^2}{Et} - \beta e^{-\beta x}[(C_1 - C_2) \cos \beta x + (C_1 + C_2) \sin \beta x]$$

(5.20)

For the lower edge at $x = 0$ these displacements must be identically equal to zero and therefore the boundary conditions are

$$u_{z(x=0)} = \gamma \frac{a^2 l}{Et} + C_1 = 0$$

$$\varphi_{x(x=0)} = -\gamma \frac{a^2}{Et} - \beta(C_1 - C_2) = 0$$

(5.21)

The solution for the constants is

$$C_1 = -\gamma \frac{a^2 l}{Et}$$

$$C_2 = -\gamma \frac{a^2 l}{Et}\left(1 - \frac{1}{\beta l}\right)$$

(5.22)

Combining the membrane solution and the edge disturbance, we obtain the circumferential membrane force $n_{\theta\theta}$ and the bending moment m_{xx}:

$$n_{\theta\theta} = \gamma a l \left\{\left(1 - \frac{x}{l}\right) - e^{-\beta x}\left[\cos \beta x + \left(1 - \frac{1}{\beta l}\right)\sin \beta x\right]\right\}$$

$$m_{xx} = -\gamma \frac{l}{2\beta^2} e^{-\beta x}\left[\left(1 - \frac{1}{\beta l}\right)\cos \beta x - \sin \beta x\right]$$

(5.23)

These quantities and the approximate values at characteristic points are shown in Fig. 5.6 for a cylindrical shell. The lateral contraction of the material is not accounted for, as we have chosen Poisson's ratio zero. We take the density of water $\gamma = 10$ kN/m^3 and choose the dimensions as $a = 3$ m, $t = 0.3$ m and $l = 4$ m. The straight line in the diagram of $n_{\theta\theta}$ represents the inhomogeneous solution and thus also represents the membrane response. The plot of $n_{\theta\theta}$ has the same course as the plot of the normal displacement u_z. The thickness-to-radius ratio, which is equal to $t/a = 1/10$, is deliberately chosen large to show that, even for this thick shell, the influence length of the edge disturbance for the axisymmetric behaviour is actually very short. Equation (5.14) yields an influence length of about $0.8\, a = 2.4$ m, which is smaller than the height of the tank ($l = 4$ m). This means that in almost every case the influence of one edge on the other will be negligible. Furthermore, it means that the inhomogeneous solution describes the global behaviour of the shell and that a bending field disturbs this global behaviour of the shell only over a relatively short section of the shell. As stated before, the inhomogeneous solution is equal to the membrane solution. To illustrate the fact that the edge disturbance is indeed very short for a thin shell under axisymmetric

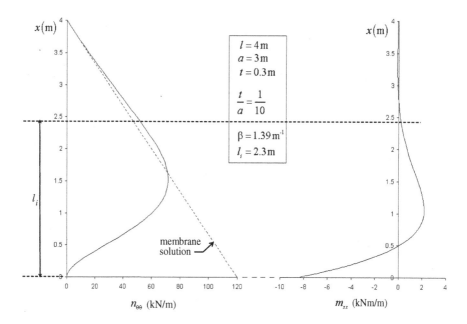

Fig. 5.6 Tank wall ($a = 10\,t$) rigidly connected to a thick bottom plate

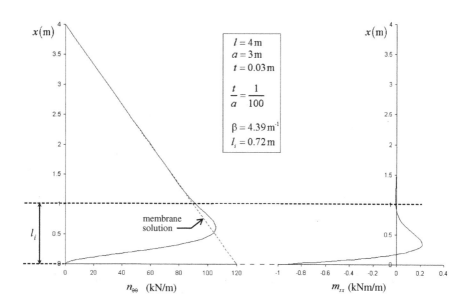

Fig. 5.7 Tank wall ($a = 100t$) rigidly connected to a thick bottom plate

loading, we repeat the calculation for a tank with the same length and radius, but with a thickness $t = 0.03$ m, see Fig. 5.7. So the thickness-to-radius ratio is equal to the value $t/a = 1/100$ and the disturbance therefore influences a much shorter length of the shell (the influence length is 0.25 $a = 1$ m). This results in a higher peak of the circumferential membrane force $n_{\theta\theta}$, but greatly reduces the peak value of the bending moment m_{xx}.

5.4 Solution for a Long Shell Subject to Edge Loads

In the water tank example of the previous section, we have implicitly seen that a transverse edge load and edge moment was applied at the base of the tank in order to compensate membrane displacements. In the present section, we will elaborate on the relationship between these edge loads and the corresponding edge displacements. We do this for shells that are longer than the influence length, and we consider the edge $x = 0$ and start from the solution as given in Eq. (5.11). Only the first term of the solution is needed:

$$u_z(x) = e^{-\beta x}[A_1 \cos \beta x + A_2 \sin \beta x] \qquad (5.24)$$

It is convenient to reshape the solution in Eq. (5.24):

$$u_z(x) = C e^{-\beta x} \sin(\beta x + \psi) \qquad (5.25)$$

Without loss of generality we have replaced the two constants A_1 and A_2 by two new constants C and the phase angle ψ. The reader may prove this by substitution of $C^2 = A_1^2 + A_2^2$ and $\tan \psi = A_1/A_2$. This presentation of the solution Eq. (5.24) is convenient for simple cases for which the phase angle can be determined immediately from the boundary conditions and Eq. (5.25) has a pleasant differentiation property. If we differentiate the function, the argument of the sine decreases by $\pi/4$ and the function gets multiplied by $-\sqrt{2}\beta$. This is the case again for the second derivative, again for the third, and again for the fourth. After four differentiations the sine argument has increased by π, which means that the original sine has changed sign, and the multiplication factor has become $\left(-\sqrt{2}\beta\right)^4 = 4\beta^4$. Substitution of the starting function $C e^{-\beta x} \sin(\beta x + \psi)$ of u_z and the fourth derivative $-4\beta^4 C e^{-\beta x} \sin(\beta x + \psi)$ in the differential equation (5.7) shows that this equation is perfectly satisfied. The expressions for the rotation ($\varphi_x = -dw/dx$), the bending moment ($m_{xx} = -D\, d^2w/dx^2$), the transverse shear force ($v_x = -D_b\, d^3w/dx^3$) and the membrane force ($n_{\theta\theta} = Et\, u_z/a^2$) become

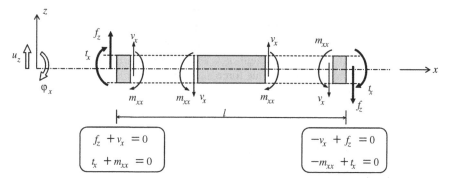

Fig. 5.8 Positive direction of the quantities of interest on a shell wall

$$u_z(x) = Ce^{-\beta x} \sin(\beta x + \psi)$$
$$\varphi_x(x) = \sqrt{2}\beta Ce^{-\beta x} \sin(\beta x + \psi - \pi/4)$$
$$m_{xx} = -D\big(2\beta^2 Ce^{-\beta x} \sin(\beta x + \psi - \pi/2)\big)$$
$$v_x = -D\Big(-2\sqrt{2}\beta^3 Ce^{-\beta x} \sin(\beta x + \psi - 3\pi/4)\Big)$$
$$n_{\theta\theta} = \frac{Et}{a} Ce^{-\beta x} \sin(\beta x + \psi)$$

(5.26)

Figure 5.8 shows a shell wall of length l. The positive directions of the displacement, rotation, forces and moments are shown. The direction of the edge loads, the distributed edge load f_z and the distributed edge torque t_x, correspond with the displacement and rotation, respectively. In the shell wall, bending moments m_{xx} and shear forces v_x occur. Note that different sign conventions apply for the external edge loads and the internal bending moment and shear force. The external loads f_z and t_x have signs as the displacement u_z and rotation φ_x.

The semi-infinite cylinder is able to withstand two axisymmetric edge loads at $x = 0$. These cases are shown in Fig. 5.9 and will be considered separately.

5.4.1 Edge Force

We find the boundary condition by considering a small beam part at each end of the shell. For the case in Fig. 5.9a the homogeneous boundary condition is

$$v_{x(x=0)} = -f_z; \quad m_{xx(x=0)} = 0 \tag{5.27}$$

We derive directly the phase angle from the expression for m_{xx} in (5.27), yielding

$$\sin\left(\psi - \frac{\pi}{2}\right) = 0 \quad \Rightarrow \quad \psi = \frac{\pi}{2} \tag{5.28}$$

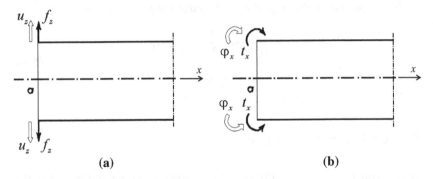

Fig. 5.9 Semi-infinite cylinder with an edge load f_z in (**a**) and an edge torque t_x in (**b**)

The expression for the constant C is derived by the expression for v_x in (5.26) for the boundary condition (5.27) with the phase angle (5.28); this leads to

$$2\sqrt{2}D\,\beta^3 C \sin\left(-\frac{\pi}{4}\right) = -f_z \quad \Rightarrow \quad C = \frac{f_z}{2D\,\beta^3} \tag{5.29}$$

The displacements of the edge due to the edge force f_z thus become:

$$u_{z\,(x=0)} = \frac{f_z}{2D\beta^3}\sin\frac{\pi}{2} = \frac{1}{2D\beta^3}f_z$$

$$\varphi_{x\,(x=0)} = \beta\sqrt{2}\frac{f_z}{2D\beta^3}\sin\frac{\pi}{4} = \frac{1}{2D\beta^2}f_z \tag{5.30}$$

5.4.2 Edge Torque

For the case in Fig. 5.9b the homogeneous boundary condition is

$$v_{x(x=0)} = 0; \quad m_{xx(x=0)} = -t_x \tag{5.31}$$

Directly from the expression for v_x in Eq. (5.26), the phase angle is derived, yielding

$$\sin\left(\psi - \frac{3\pi}{4}\right) = 0 \quad \Rightarrow \quad \psi = \frac{3\pi}{4} \tag{5.32}$$

The expression for the constant C in Eq. (5.26) is derived by the expression for m_{xx} for the boundary condition in Eq. (5.31) with the phase angle of Eq. (5.32), which results in

$$-2D\beta^2 C \sin\left(\frac{\pi}{4}\right) = -t_x \quad \Rightarrow \quad C = \frac{t_x}{\sqrt{2}D\beta^2} \tag{5.33}$$

The displacements of the edge due to the edge torque t_x thus become

$$u_z = \frac{t_x}{\sqrt{2}D\beta^2}\sin\frac{3\pi}{4} = \frac{1}{2D\beta^2}t_x$$

$$\varphi_x = \beta\sqrt{2}\frac{t_x}{\sqrt{2}D\beta^2}\sin\frac{\pi}{2} = \frac{1}{D\beta}t_x \qquad (5.34)$$

5.4.3 Edge Force and Torque

We use the general solution of Eq. (5.26) to determine the flexibility matrix that relates the displacement u_z and rotation φ_x with the edge load f_z and edge torque t_x. The superposition of Eqs. (5.30) and (5.34), gives the combined action of the edge loads as shown in Fig. 5.10. Presented symbolically as $\mathbf{Ff} = \mathbf{u}$, this is

$$\frac{1}{D}\begin{bmatrix} \dfrac{1}{2\beta^3} & \dfrac{1}{2\beta^2} \\ \dfrac{1}{2\beta^2} & \dfrac{1}{\beta} \end{bmatrix}\begin{bmatrix} f_z \\ t_x \end{bmatrix} = \begin{bmatrix} u_z \\ \varphi_x \end{bmatrix} \qquad (5.35)$$

Matrix $\mathbf{F}$ is the *flexibility matrix*. This holds for an edge at the base of the shell ($x = 0$). For the edge at the top of the shell ($x = l$), see Fig. 5.11, the flexibility matrix is

$$\frac{1}{D}\begin{bmatrix} \dfrac{1}{2\beta^3} & -\dfrac{1}{2\beta^2} \\ -\dfrac{1}{2\beta^2} & \dfrac{1}{\beta} \end{bmatrix}\begin{bmatrix} f_z \\ t_x \end{bmatrix} = \begin{bmatrix} u_z \\ \varphi_x \end{bmatrix} \qquad (5.36)$$

In accordance with Maxwell's reciprocal theorem, the matrix $\mathbf{F}$ is symmetrical and positive definite in both cases.

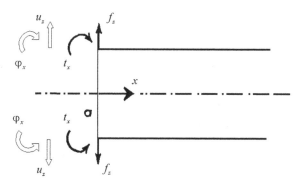

Fig. 5.10   Semi-infinite cylinder with an edge load f_z and torque t_x

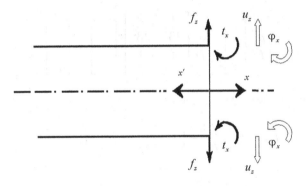

Fig. 5.11 Semi-infinite cylinder with an edge load f_z and an edge torque t_x

5.5 Reconsidering the Water Tank by Force Method

This example is actually the same as in Sect. 5.3 but now we will use the *force method* to compute the bending action at the edge; the formulation (5.35) is used. We need the membrane solution due to the axisymmetric water for the displacement u_z and the rotation $\varphi_x = -du_z/dx$:

$$u_{z,m} = \gamma \frac{a^2}{Et}(l - x)$$
$$\varphi_{x,m} = \gamma \frac{a^2}{Et}$$

(5.37)

The total displacements $\mathbf{u_t}$ of an edge is the sum of the membrane displacements $\mathbf{u_m}$ and the bending displacements $\mathbf{u_b}$. Therefore continuity of displacements demands

$$\mathbf{u_t} = \mathbf{u_m} + \mathbf{u_b}$$

(5.38)

For the lower edge at $x = 0$, which is clamped, this yields

$$\begin{bmatrix} u_z \\ \varphi_x \end{bmatrix}^t = \begin{bmatrix} u_z \\ \varphi_x \end{bmatrix}^m + \begin{bmatrix} u_z \\ \varphi_x \end{bmatrix}^b = \begin{bmatrix} 0 \\ 0 \end{bmatrix}$$

(5.39)

The bending action is represented by the relation (5.35):

$$\frac{1}{D}\begin{bmatrix} \dfrac{1}{2\beta^3} & \dfrac{1}{2\beta^2} \\ \dfrac{1}{2\beta^2} & \dfrac{1}{\beta} \end{bmatrix}\begin{bmatrix} f_z \\ t_x \end{bmatrix} = \begin{bmatrix} u_z \\ \varphi_x \end{bmatrix}^b$$

(5.40)

By putting together the membrane action for $x = 0$ and the bending action, the following set of two equations with two unknown edge loads is obtained:

$$\gamma \frac{a^2}{Et} \begin{bmatrix} l \\ 1 \end{bmatrix} + \frac{1}{D} \begin{bmatrix} \dfrac{1}{2\beta^3} & \dfrac{1}{2\beta^2} \\ \dfrac{1}{2\beta^2} & \dfrac{1}{\beta} \end{bmatrix} \begin{bmatrix} f_z \\ t_x \end{bmatrix} = \begin{bmatrix} 0 \\ 0 \end{bmatrix} \tag{5.41}$$

The solution of these two equations is

$$f_z = -\gamma \frac{4D\beta^3 a^2 l}{Et}\left(1 - \frac{1}{\beta l}\right) = -\gamma \frac{l}{\beta}\left(1 - \frac{1}{\beta l}\right)$$

$$\tag{5.42}$$

$$t_x = \gamma \frac{2D\beta^2 a^2 l}{Et}\left(1 - \frac{1}{\beta l}\right) = \gamma \frac{l}{2\beta^2}\left(1 - \frac{1}{\beta l}\right)$$

To obtain this result we accounted for Eq. (5.6).

5.6 Four Elementary Cases

With the flexibility relation (5.35) between the edge load vector **f** and the edge displacement vector **u**, we can analyse four elementary cases [1]. These are shown in Fig. 5.12. To exemplify the method, the solution for one elementary case (case A) is shown. The other elementary cases can be derived similarly. For these cases we give only the boundary conditions and leave it up to the reader to verify the analytical expressions and relevant values. The course of the quantities of interest is shown in Figs. 5.12 and 5.13, and the analytical expressions are tabulated in Table 5.1.

5.6.1 Elementary Case A

A cylinder of infinite length in both directions is loaded by a distributed line load $2f_0$, uniformly distributed in the circumferential direction, see Fig. 5.12. Because of symmetry considerations, this case is equal to a cylinder which starts at $x = 0$, is of infinite length in one direction, is loaded by a uniformly distributed line load f_0, and has zero rotation at that end $x = 0$. The boundary conditions at the end $x = 0$ are

$$\varphi_x = -\left(\frac{du_z}{dx}\right)_{(x=0)} = 0$$

$$\tag{5.43}$$

$$f_z = -v_{x(x=0)} = -f_0$$

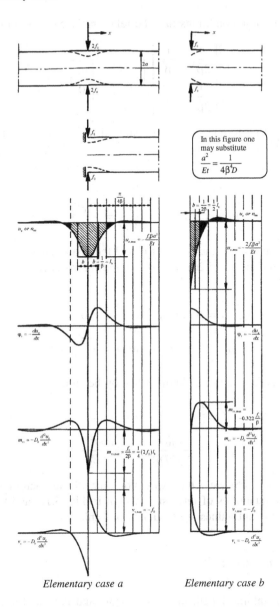

Elementary case a   Elementary case b

Fig. 5.12 Two elementary cases. Distribution of the quantities of interest

By substituting these conditions into the relation (5.35), it becomes

$$\frac{1}{D} \begin{bmatrix} \dfrac{1}{2\beta^3} & \dfrac{1}{2\beta^2} \\ \dfrac{1}{2\beta^2} & \dfrac{1}{\beta} \end{bmatrix} \begin{bmatrix} -f_0 \\ t_x \end{bmatrix} = \begin{bmatrix} u_z \\ 0 \end{bmatrix} \tag{5.44}$$

The unknown edge torque t_x and the edge displacement u_z can be obtained subsequently and become

$$t_x = -m_{xx(x=0)} = \frac{f_0}{2\beta}$$

$$u_z = u_{z(x=0)} = -\frac{f_0}{4\beta^3 D} = -\frac{f_0 \beta a^2}{Et} \tag{5.45}$$

For this elementary case, the constant C and the phase angle ψ can be directly determined with the aid of (5.26). For $x = 0$ and the boundary conditions, these expressions read

$$\varphi_{x(x=0)} = \beta\sqrt{2}C \sin\left(\psi - \frac{\pi}{4}\right) = 0$$

$$v_{x(x=0)} = 2\sqrt{2}D\beta^3 C \sin\left(\psi - \frac{3\pi}{4}\right) = f_0 \tag{5.46}$$

The expressions for the two unknowns hereby become

$$\psi = \frac{\pi}{4}$$

$$C = -\frac{f_0}{2\sqrt{2}D\beta^3} = -\frac{f_0 \beta a^2 \sqrt{2}}{Et} \tag{5.47}$$

The expressions (5.26) yield the expressions for all the quantities of interest. The course of the quantities of interest is shown in Fig. 5.12, and the analytical expressions and relevant values are tabulated in Table 5.1.

5.6.2 Elementary Case B

The cylinder is of infinite length in one direction, and is loaded at the end by a uniformly distributed line load f_0 in circumferential direction. The boundary conditions at the end $x = 0$ are

$$t_x = -m_{xx(x=0)} = 0$$

$$f_z = -v_{x(x=0)} = -f_0 \tag{5.48}$$

The course of the quantities of interest is shown in Fig. 5.12, and the analytical expressions and relevant values are tabulated in Table 5.1.

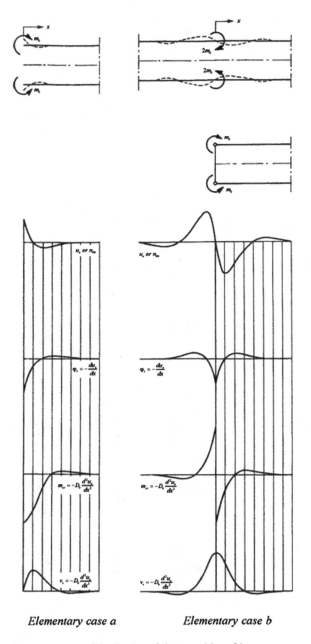

Elementary case a   Elementary case b

Fig. 5.13 Two elementary cases. Distribution of the quantities of interest

Table 5.1 Four elementary cases of the axisymmetrically loaded cylinder. In this table we may substitute $\dfrac{a^2}{Et} = \dfrac{1}{4\beta^4 D_b}$

	u_z	$u_{z,\max}$	$\varphi_x = -\dfrac{du_z}{dx}$	$\varphi_{x,\max}$	$m_{xx} = -D_b\dfrac{d^2 u_z}{dx^2}$	$m_{x,\max}$	$v_x = -D_b\dfrac{d^3 u_z}{dx^3}$	$v_{x,\max}$
Elementary case a	$-f_0\beta\sqrt{2}\dfrac{a^2}{Et}e^{-\beta x}\sin\left(\beta x+\dfrac{\pi}{4}\right)$	$-\dfrac{f_0\beta a^2}{Et}$ at $x=0$	$-2f_0\beta^2\dfrac{a^2}{Et}e^{-\beta x}\sin\beta x$	$-0.644 f_0\beta^2\dfrac{a^2}{Et}$ at $x=\dfrac{\pi}{4\beta}$	$-\dfrac{f_0}{2\beta}\sqrt{2}e^{-\beta x}\sin\left(\beta x-\dfrac{\pi}{4}\right)$	$\dfrac{f_0}{2\beta}$ at $x=0$	$f_0 e^{-\beta x}\sin\left(\beta x-\dfrac{\pi}{2}\right)$	$-f_0$ at $x=0$
Elementary case b	$-2f_0\beta^2\dfrac{a^2}{Et}e^{-\beta x}\sin\left(\beta x+\dfrac{\pi}{2}\right)$	$-2\dfrac{f_0\beta a^2}{Et}$ at $x=0$	$-2f_0\beta^2\sqrt{2}\dfrac{a^2}{Et}e^{-\beta x}\sin\left(\beta x+\dfrac{\pi}{4}\right)$	$-2f_0\beta^2\dfrac{a^2}{Et}$ at $x=0$	$-\dfrac{f_0}{\beta}e^{-\beta x}\sin\beta x$	$-0.322\dfrac{f_0}{\beta}$ at $x=\dfrac{\pi}{4\beta}$	$f_0\sqrt{2}e^{-\beta x}\sin\left(\beta x-\dfrac{\pi}{4}\right)$	$-f_0$ at $x=0$
Elementary case c	$\dfrac{m_0}{D_b\beta^2\sqrt{2}}e^{-\beta x}\sin\left(\beta x+\dfrac{3\pi}{4}\right)$	$\dfrac{m_0}{2D_b\beta^2}$ at $x=0$	$\dfrac{m_0}{D_b\beta}e^{-\beta x}\sin\left(\beta x+\dfrac{\pi}{2}\right)$	$\dfrac{m_0}{D_b\beta}$ at $x=0$	$m_0\sqrt{2}e^{-\beta x}\sin\left(\beta x+\dfrac{\pi}{4}\right)$	m_0 at $x=0$	$-2m_0\beta e^{-\beta x}\sin\beta x$	$-0.644\beta m_0$ at $x=\dfrac{\pi}{4\beta}$
Elementary case d	$\dfrac{m_0}{2D_b\beta^2}e^{-\beta x}\sin(\beta x+\pi)$	$0.161\dfrac{m_0}{D_b\beta^2}$ at $x=\dfrac{\pi}{4\beta}$	$\dfrac{m_0}{2D_b\beta}e^{-\beta x}\sin\left(\beta x+\dfrac{3\pi}{4}\right)$	$\dfrac{m_0}{2D_b\beta}$ at $x=0$	$-m_0 e^{-\beta x}\sin\left(\beta x+\dfrac{\pi}{2}\right)$	m_0 at $x=0$	$-m_0\beta\sqrt{2}e^{-\beta x}\sin\left(\beta x+\dfrac{\pi}{4}\right)$	$-m_0\beta$ at $x=0$

5.6.3 Elementary Case C

The cylinder is of infinite length in one direction and is loaded at the end by a uniformly distributed torque m_0. The boundary conditions at the end $(x = 0)$ are

$$t_x = -m_{xx(x=0)} = m_0$$
$$f_z = -v_{x(x=0)} = 0$$
$$(5.49)$$

The course of the quantities of interest is shown in Fig. 5.13, and the analytical expressions and relevant values are tabulated in Table 5.1.

5.6.4 Elementary Case D

The cylinder is of infinite length in both directions and is locally loaded by a torque $2m_0$. Because of symmetry considerations we can consider an infinitely long cylinder in one direction, which is pin-supported and loaded by a torque m_0. The boundary conditions at the end $x = 0$ are:

$$u_z = u_{z(x=0)} = 0$$
$$t_x = -m_{xx(x=0)} = m_0$$
$$(5.50)$$

The course of the quantities of interest is shown in Fig. 5.13, and the analytical expressions and relevant values are tabulated in Table 5.1.

5.7 Concluding Remarks

It is shown in this chapter that the description of the combined stretching and bending behaviour of a thin shell under axisymmetric loading is adequately achieved by combining the membrane behaviour of the shell with the bending behaviour of a flat plate. The bending theory leads to a single differential equation of the fourth order for the normal displacement.

In fact the membrane solution is the inhomogeneous solution of the differential equation. If the membrane displacements at the boundaries are not consistent with the actual boundary conditions, the solution to the homogeneous bending equation can be used as an edge disturbance to the membrane behaviour. We have shown that, for the axisymmetric case, the influence of this edge disturbance is local and that the influence length is often much smaller than the length of the shell structure. To simplify the calculation procedure the deformation of an edge of a circular cylindrical shell due to axisymmetric edge loads is derived.

Reference

1. Bouma AL et al (1958) Edge disturbances in axi-symmetrically loaded shells of rotation. IBC announcements, Institute TNO for Building Materials and Building Structures, Delft (in Dutch)

Part II
Roof Structures

Chapter 6
Donnell Bending Theory for Shallow Shells

In this chapter we will extend the membrane theory for shells of arbitrary curvatures, as presented in Chap. 3, to a theory in which we account for both membrane and bending action. This theory, developed by Donnell [1], is applicable to *shallow* shells like roof shells. The *Donnell theory* is not sufficiently accurate for circular cylindrical shells like chimneys and storage tanks. These structures are *deep* shells instead of shallow ones. Hereafter, in Chap. 9, we will present a more rigorous theory for this type of shell.

6.1 Introduction

Chapter 3 deals with the membrane theory for shells of arbitrary curvature. The purpose of this chapter is to derive an appropriate bending theory for such shells. In the bending theory, a coupling occurs between the membrane action and the bending action. Different from the edge disturbance theory in Chap. 5, bending moments may occur now over the total area of the shell surface. In the present chapter we derive the differential equation for shallow shells, which is applicable to shell roof structures, see Fig. 6.1. In this theory, the expressions for the change of curvature due to bending moments are borrowed from the theory of flat plates. Moreover, the difference between the global axes x, y, z to describe the geometry of the shell and the local axes x, y, z in the shell surface with z normal to the surface is neglected. This is permitted if the slope of the roof is sufficiently small. As done in the previous chapter for edge disturbance problems, the deformation due to transverse shear forces is neglected in this theory. Summing up, membrane forces n_{xx}, n_{yy}, n_{xy} and bending/twisting moments m_{xx}, m_{yy}, m_{xy} play a role in the derivation of the differential equation.

J. Blaauwendraad and J. H. Hoefakker, *Structural Shell Analysis*,
Solid Mechanics and Its Applications 200, DOI: 10.1007/978-94-007-6701-0_6,
© Springer Science+Business Media Dordrecht 2014

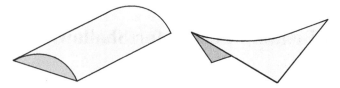

Fig. 6.1 Shallow roof shells

Notation and Sign Convention

The notation and sign convention for stresses and membrane forces of Chap. 2 applies again. A similar notation with two indices is used for the bending moments m_{xx} and m_{yy} and the twisting moment m_{xy}. The sign convention is as follows. A bending moment is positive if the bending stress at the positive z-side of the middle plane is positive. The twisting moment is positive if the shear stress at the positive z-side of the middle plane is positive. Transverse shear forces are written as v_x and v_y, respectively. The subscript indicates the normal to the face on which the shear force occurs. The force is positive if acting in the positive z-direction on a face with positive normal. Figure 6.2 shows positive moments and shear forces.

The displacements can vary in all directions. As a consequence, the displacement u_θ, which is zero in case of axisymmetric behaviour of circular cylindrical shells, will be unequal to zero. Therefore, all strains ε_{xx}, ε_{yy} and γ_{xy} play a role. The shell is bent in x- and y-direction. So, in general change of curvatures κ_{xx} and κ_{yy} will occur. Subsequently, the surface is also twisted and the analysis will also involve the torsion deformation, which we denote by ρ_{xy}. The strains and curvatures correlate with membrane forces n_{xx}, n_{yy}, n_{xy} and bending/twisting moments m_{xx}, m_{yy}, m_{xy}. We use the following vectors

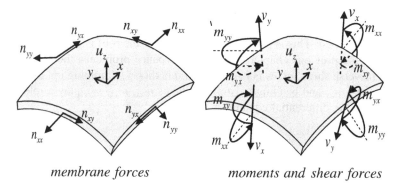

membrane forces   *moments and shear forces*

Fig. 6.2 Forces and moments on a shell element of arbitrary curvature

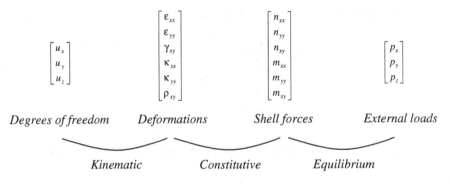

Fig. 6.3   Scheme of relationships for a shell

$$\mathbf{u} = \begin{bmatrix} u_x & u_y & u_z \end{bmatrix}^{\mathrm{T}}$$
$$\mathbf{e} = \begin{bmatrix} \varepsilon_{xx} & \varepsilon_{yy} & \gamma_{xy} & \kappa_{xx} & \kappa_{yy} & \rho_{xy} \end{bmatrix}^{\mathrm{T}}$$
$$\mathbf{s} = \begin{bmatrix} n_{xx} & n_{yy} & n_{xy} & m_{xx} & m_{yy} & m_{xy} \end{bmatrix}^{\mathrm{T}} \tag{6.1}$$
$$\mathbf{p} = \begin{bmatrix} p_x & p_y & p_z \end{bmatrix}^{\mathrm{T}}$$

Hereafter we call $\mathbf{u}$ the displacement vector, $\mathbf{e}$ the deformation vector, $\mathbf{s}$ the vector of shell forces (membrane forces and bending/twisting moments) and $\mathbf{p}$ the load vector. The scheme of relationships of Fig. 6.3 applies.

6.2 Kinematic Relation

For the membrane strains the relation (3.14) of the membrane behaviour is used:

$$\varepsilon_{xx} = \frac{\partial u_x}{\partial x} - k_x u_z$$
$$\varepsilon_{yy} = \frac{\partial u_y}{\partial y} - k_y u_z \tag{6.2}$$
$$\gamma_{xy} = \frac{\partial u_x}{\partial y} + \frac{\partial u_y}{\partial x} - 2k_{xy} u_z$$

The rotations of the bending behaviour of a flat plate applied to a thin shell of arbitrary curvature with the infinitesimal length of arcs dx and dy are

$$\varphi_x = -\frac{\partial u_z}{\partial x}$$
$$\varphi_y = -\frac{\partial u_z}{\partial y} \tag{6.3}$$

Hereby, the bending deformations become

$$\kappa_{xx} = \frac{\partial \varphi_x}{\partial x} = -\frac{\partial^2 u_z}{\partial x^2}$$
$$\kappa_{yy} = \frac{\partial \varphi_y}{\partial y} = -\frac{\partial^2 u_z}{\partial y^2} \tag{6.4}$$
$$\rho_{xy} = \frac{\partial \varphi_x}{\partial y} + \frac{\partial \varphi_y}{\partial x} = -2\frac{\partial^2 u_z}{\partial x \partial y}$$

Consequently, the kinematic relation, symbolically presented as $\mathbf{e} = \mathbf{B}\mathbf{u}$, reads:

$$
\begin{bmatrix} \varepsilon_{xx} \\ \varepsilon_{yy} \\ \gamma_{xy} \\ \kappa_{xx} \\ \kappa_{yy} \\ \rho_{xy} \end{bmatrix}
=
\begin{bmatrix}
\frac{\partial}{\partial x} & 0 & -k_x \\
0 & \frac{\partial}{\partial y} & -k_y \\
\frac{\partial}{\partial y} & \frac{\partial}{\partial x} & -2k_{xy} \\
0 & 0 & -\frac{\partial^2}{\partial x^2} \\
0 & 0 & -\frac{\partial^2}{\partial y^2} \\
0 & 0 & -2\frac{\partial^2}{\partial x \partial y}
\end{bmatrix}
\begin{bmatrix} u_x \\ u_y \\ u_z \end{bmatrix}. \tag{6.5}
$$

6.3 Constitutive Relation

Symbolically presented as $\mathbf{s} = \mathbf{D}\,\mathbf{e}$, this constitutive relation between the shell forces vector and the deformation vector become

$$
\begin{bmatrix} n_{xx} \\ n_{yy} \\ n_{xy} \\ m_{xx} \\ m_{yy} \\ m_{xy} \end{bmatrix}
=
\begin{bmatrix}
D_m & vD_m & 0 & 0 & 0 & 0 \\
vD_m & D_m & 0 & 0 & 0 & 0 \\
0 & 0 & D_m\left(\frac{1-v}{2}\right) & 0 & 0 & 0 \\
0 & 0 & 0 & D_b & vD_b & 0 \\
0 & 0 & 0 & vD_b & D_b & 0 \\
0 & 0 & 0 & 0 & 0 & D_b\left(\frac{1-v}{2}\right)
\end{bmatrix}
\begin{bmatrix} \varepsilon_{xx} \\ \varepsilon_{yy} \\ \gamma_{xy} \\ \kappa_{xx} \\ \kappa_{yy} \\ \rho_{xy} \end{bmatrix} \tag{6.6}
$$

$\mathbf{D}$ is the rigidity matrix. The membrane rigidity D_m and the flexural rigidity D_b are

$$D_m = \frac{Et}{(1 - v^2)}, \qquad D_b = \frac{Et^3}{12(1 - v^2)}. \tag{6.7}$$

6.4 Equilibrium Relation

Corresponding to the analysis of Chap. 4, the equilibrium of forces in the tangential directions is fully governed by the membrane behaviour. In the normal direction, not only the membrane behaviour, but also the bending behaviour is taken into account. The transverse shear forces do contribute to the out-of-plane equilibrium, but the in-plane equilibrium is solely described by the membrane forces. By adding the contribution of the transverse shear forces to the normal equilibrium of the membrane relation Eqs. (3.27) and (3.29) we find

$$\frac{\partial n_{xx}}{\partial x} + \frac{\partial n_{xy}}{\partial y} + p_x = 0$$

$$\frac{\partial n_{yy}}{\partial y} + \frac{\partial n_{xy}}{\partial x} + p_y = 0 \tag{6.8}$$

$$\frac{\partial v_x}{\partial x} + \frac{\partial v_y}{\partial y} + k_x n_{xx} + k_y n_{yy} + 2k_{xy} n_{xy} + p_z = 0$$

According to the bending action of a flat plate, the equilibrium of moments with respect to the y-direction and to the x-direction yields the following equations for the transverse shear forces:

$$v_x = \frac{\partial m_{xx}}{\partial x} + \frac{\partial m_{xy}}{\partial y}$$

$$v_y = \frac{\partial m_{yy}}{\partial y} + \frac{\partial m_{xy}}{\partial x} \tag{6.9}$$

By substitution of these shear forces, Eq. (6.8) becomes the equilibrium relation:

$$
\begin{bmatrix}
-\dfrac{\partial}{\partial x} & 0 & -\dfrac{\partial}{\partial y} & 0 & 0 & 0 \\[2mm]
0 & -\dfrac{\partial}{\partial y} & -\dfrac{\partial}{\partial x} & 0 & 0 & 0 \\[2mm]
-k_x & -k_y & -2k_{xy} & -\dfrac{\partial^2}{\partial x^2} & -\dfrac{\partial^2}{\partial y^2} & -2\dfrac{\partial^2}{\partial x \partial y}
\end{bmatrix}
\begin{bmatrix}
n_{xx} \\ n_{yy} \\ n_{xy} \\ m_{xx} \\ m_{yy} \\ m_{xy}
\end{bmatrix}
=
\begin{bmatrix}
p_x \\ p_y \\ p_z
\end{bmatrix}
\tag{6.10}
$$

Symbolically presented as $\mathbf{B}^*\mathbf{s} = \mathbf{p}$, the differential operator matrix $\mathbf{B}^*$ is the adjoint of the differential operator matrix $\mathbf{B}$ in Eq. (6.5) for the kinematic relation.

6.5 Differential Equation for One Displacement

In classical plate theories, a different approach is often followed for plates loaded in-plane (membrane forces) and plates loaded out-of-plane (moments and transverse shear forces). In the first category the *force method* is followed, and in the second the *stiffness method*. In the force method, we first solve the equilibrium equations by introduction of the Airy stress function Φ and end with the kinematic equation, from which a compatibility condition is derived. This condition leads to a biharmonic differential equation for the stress function. In the second category, the stiffness method, the derivation starts with the kinematic equations and proceeds to the equilibrium equations, ending up again with a biharmonic equation, now for the normal displacement u_z. In the shallow shell, the two plate actions, membrane and bending, are coupled. Yet it is convenient to work along similar lines as for plates. We will use the force method for the membrane action and the stiffness method procedure for the bending action. We end up with two-fourth-order partial differential equations with two unknowns, the stress function Φ and the normal displacement u_z. From these two-fourth-order equations, one eight-order partial differential equation in the normal displacement can be derived.

In the remainder of the section, we will exclude the contributions of the loads p_x, p_y and p_z. The response to the loads will often not be evaluated via the differential equation, but most likely be obtained from the membrane solution or equilibrium equations. Then, the differential equation is just needed to satisfy the boundary conditions in case of membrane incompatible edges. Therefore, the load to be considered for the differential equation consists of edge forces and moments. Distributed loads do not play any role. From here on we put $p_x = 0$, $p_y = 0$ and $p_z = 0$.

6.5.1 In-Plane State

For the in-plane state we apply the force method, and start with solving the first two equations in Eq. (6.10). We introduce the Airy stress function Φ, so that

$$n_{xx} = \frac{\partial^2 \Phi}{\partial y^2}, \; n_{yy} = \frac{\partial^2 \Phi}{\partial x^2}, \; n_{xy} = -\frac{\partial^2 \Phi}{\partial x \partial y} \tag{6.11}$$

By this definition, the two equations are identically satisfied for zero loads $p_x = 0$ and $p_y = 0$. The second step in the procedure concerns the constitutive relations. In the force method these are used in the inverse way:

$$\varepsilon_{xx} = \frac{1}{D_m(1 - v^2)}\left(n_{xx} - v n_{yy}\right)$$

$$\varepsilon_{yy} = \frac{1}{D_m(1 - v^2)}\left(-v n_{xx} + n_{yy}\right) \tag{6.12}$$

$$\gamma_{xy} = \frac{2(1 + v)}{D_m(1 - v^2)}n_{xy}$$

In the third step, we operate on the kinematic relation for the membrane state. We eliminate the in-plane displacements u_x and u_y from the first three equations of kinematic relation (6.5). From these equations we derive

$$\frac{\partial^2 \varepsilon_{xx}}{\partial y^2} + \frac{\partial^2 \varepsilon_{yy}}{\partial x^2} = \frac{\partial^3 u_x}{\partial x \partial y^2} + \frac{\partial^3 u_y}{\partial x^2 \partial y} - k_x \frac{\partial^2 u_z}{\partial y^2} - k_y \frac{\partial^2 u_z}{\partial x^2}$$

$$\frac{\partial^2 \gamma_{xy}}{\partial x \partial y} = \frac{\partial^3 u_x}{\partial x \partial y^2} + \frac{\partial^3 u_y}{\partial x^2 \partial y} - 2k_{xy} \frac{\partial^2 u_z}{\partial x \partial y} \tag{6.13}$$

Taking the difference of these expressions eliminates the displacements u_x and u_y:

$$\frac{\partial^2 \varepsilon_{xx}}{\partial y^2} - \frac{\partial^2 \gamma_{xy}}{\partial x \partial y} + \frac{\partial^2 \varepsilon_{yy}}{\partial x^2} = -k_x \frac{\partial^2 u_z}{\partial y^2} + 2k_{xy} \frac{\partial^2 u_z}{\partial x \partial y} - k_y \frac{\partial^2 u_z}{\partial x^2} \tag{6.14}$$

This differential equation is not only a transitional product of our manipulation, but also has a physical meaning. By using the kinematic relation for the changes of curvature in Eq. (6.5), the result is rewritten as

$$\frac{\partial^2 \varepsilon_{xx}}{\partial y^2} - \frac{\partial^2 \gamma_{xy}}{\partial x \partial y} + \frac{\partial^2 \varepsilon_{yy}}{\partial x^2} = k_y \kappa_{xx} - k_{xy} \kappa_{xy} + k_x \kappa_{yy} \tag{6.15}$$

This equation is the *compatibility condition*. Strains and curvatures can not change freely; they must change so that Eq. (6.15) is satisfied. The three Eqs. (6.12), (6.14) and (6.15) are the basis for the in-plane differential equation. We substitute Eq. (6.12) into (6.14), and the new Eq. (6.14) into (6.15). This leads to the fourth-order differential equation

$$\frac{\partial^4 \Phi}{\partial x^4} + 2\frac{\partial^4 \Phi}{\partial x^2 \partial y^2} + \frac{\partial^4 \Phi}{\partial y^4} + D_m(1 - v^2)\left(k_x \frac{\partial^2 u_z}{\partial y^2} - 2k_{xy} \frac{\partial^2 u_z}{\partial x \partial y} + k_y \frac{\partial^2 u_z}{\partial x^2} \right) = 0 \tag{6.16}$$

For zero k_x, k_y and k_{xy}, the differential equation reduces to the well-known biharmonic differential equation of Airy.

6.5.2 Out-of-Plane State

For the out-of-plane state we start with the kinematic relation. We substitute the last three kinematic equations for κ_{xx}, κ_{yy} and ρ_{xy} of Eq. (6.5) into the constitutive relation Eq. (6.6) for the moments m_{xx}, m_{yy} and m_{xy}, obtaining

$$m_{xx} = -D_b\left(\frac{\partial^2 u_z}{\partial x^2} + v\frac{\partial^2 u_z}{\partial y^2} \right)$$

$$m_{yy} = -D_b\left(v\frac{\partial^2 u_z}{\partial x^2} + \frac{\partial^2 u_z}{\partial y^2} \right) \tag{6.17}$$

$$m_{xy} = -D_b(1 - v)\frac{\partial^2 u_z}{\partial x \partial y}$$

We now proceed to the third equilibrium equation of Eq. (6.10). By substitution of Eq. (6.17), accounting for Eq. (6.11), we obtain a second fourth-order differential equation:

$$-k_x \frac{\partial^2 \Phi}{\partial y^2} + 2k_{xy} \frac{\partial^2 \Phi}{\partial x \partial y} - k_y \frac{\partial^2 \Phi}{\partial x^2} + D_b \left(\frac{\partial^4 u_z}{\partial x^4} + 2 \frac{\partial^4 u_z}{\partial x^2 \partial y^2} + \frac{\partial^4 u_z}{\partial y^4} \right) = 0 \qquad (6.18)$$

For zero k_x, k_y and k_{xy}, the differential equation reduces to the well-known biharmonic differential equation of the Kirchhoff plate theory for zero load p_z.

6.5.3 Coupled States

The differential equations (6.16) and (6.18) are coupled equations for the two unknowns Φ and u_z. In this subsection we will replace them by a single differential equation for u_z. For that purpose, it is convenient to rearrange the equations, for which we introduce the differential operator Γ and the Laplacian Δ:

$$\Gamma = k_x \frac{\partial^2}{\partial y^2} - 2k_{xy} \frac{\partial^2}{\partial x \partial y} + k_y \frac{\partial^2}{\partial x^2}, \quad \Delta = \frac{\partial^2}{\partial x^2} + \frac{\partial^2}{\partial y^2} \qquad (6.19)$$

By using the definitions in Eq. (6.19), we can rewrite Eqs. (6.16) and (6.18) in the form

$$\begin{aligned} -\Gamma \Phi + D_b \Delta u_z &= 0 \\ \Delta \Delta \Phi + D_m (1 - v^2) \Gamma u_z &= 0 \end{aligned} \qquad (6.20)$$

We consider shells of constant curvatures k_x, k_y and k_{xy}. Then, the operators Γ and Δ commute, and thus $\Delta \Gamma = \Gamma \Delta$. This allows us to obtain the wanted single differential equation for the displacement u_z. The first equation in (6.20), the 'equilibrium differential equation', is multiplied by $\Delta \Delta$, and the second equation, the 'kinematic differential equation', is multiplied by Γ. By subsequently eliminating the stress function we find the single differential equation

$$D_b \Delta \Delta \Delta \Delta u_z + D_m (1 - v^2) \Gamma^2 u_z = 0 \qquad (6.21)$$

This elegant eight-order partial differential equation will be the starting point for the analysis of roof shells in the Chaps. 7 and 8. If the homogeneous solution to the differential equation is determined, the stress function Φ can be computed by Eq. (6.20) and thereupon the other quantities can be determined.

At the start of Sect. 6.5 we put p_x, p_y and p_z to zero. If not all distributed loads are put to zero, but we would consider nonzero loads p_z normal to the surface, the differential equation would become

$$D_b \Delta\Delta\Delta\Delta u_z + D_m(1 - v^2)\Gamma^2 u_z = \Delta\Delta p_z. \tag{6.22}$$

Now both a homogenous and inhomogeneous solution must be obtained. In general, the inhomogeneous solution is equal to the membrane solution. As stated before, in this book we will make use of this knowledge, and apply the homogeneous solution of Eq. (6.21) to match the boundary conditions.

In the limit case that the shell degenerates to a flat plate, the curvatures k_x, k_y and k_{xy} are zero. Then, according to Eq. (6.17), the operator Γ is zero as well. Now differential equation (6.22) reduces to the well-known biharmonic equation for plate bending,

$$D_b \Delta\Delta u_z = p_z \tag{6.23}$$

6.6 Boundary Conditions

Solving the eight-order differential equation presupposes that we apply 16 boundary conditions, four at both straight edges and four at both curved edges of the cylindrical shell. Two conditions occur at each edge for the membrane state and two for the bending state. Boundaries can be put either for displacements or for stress resultants.

For the boundary conditions corresponding with bending, we refer to *Kirchhoff theory* for thin plate bending, expecting that the reader is familiar with it. In this theory the normal edge displacement may be specified, or alternatively, the *Kirchhoff shear force*. The Kirchhoff shear forces $*v_x$ and $*v_y$ are a combination of the real shear force and the twisting moment.

$$\begin{aligned}
*v_x &= v_x + \frac{\partial m_{\theta x}}{a\,\partial\theta} \\
*v_\theta &= v_\theta + \frac{\partial m_{x\theta}}{\partial x}
\end{aligned} \tag{6.24}$$

Straight edges
At straight edges we may specify

- either u_x or $n_{\theta x}$ $\hspace{4cm}$ (6.25)

- either u_θ or $n_{\theta\theta}$ $\hspace{4cm}$ (6.26)

- either u_z or $*v_\theta$ $\hspace{4cm}$ (6.27)

- either φ_θ or $m_{\theta\theta}$ $\hspace{4cm}$ (6.28)

Curved edges
At curved edges we may specify

- either u_x or n_{xx} (6.29)

- either u_θ or $n_{x\theta}$ (6.30)

- either u_z or $^*v_x$ (6.31)

- either φ_x or m_{xx} (6.32)

Reference

1. Donnell LH (1933) Stability of thin-walled tubes under torsion. NACA Report No. 479

Chapter 7
Circular Cylindrical Roof

In this chapter, we study the bending behaviour of a circular cylindrical roof shell under asymmetric loading. The shell is part of a full circular cylinder, and is supported at the curved edges by a diaphragm (tympan) which is infinitely rigid in its plane and perfectly flexible out-of-plane. The shell is pin-connected to the diaphragm. The roof structure can be considered as a shallow shell, so the Donnell theory [1] of Chap. 6 is applicable. The theory presented here was worked out by Bouma et al. Bouma was the leader of a project team in The Netherlands, in which also Loof and Van Koten contributed, making the demanding theory accessible to practitioners [2, 3]. Von Karman et al. extended Donnell's theory to the study of buckling [4] and Jenkins [5] was the first to apply it to circular cylindrical roofs (applying matrix analysis). Because of the respective contributions of Donnell, Karman and Jenkins, the Dutch team used to refer to the theory as the DKJ-method.

7.1 Introduction

Different from the water tank problem in Chap. 5, where the disturbance was at the base circular edge, we now have a disturbance starting from the straight edges (see Fig. 7.1), and boundary conditions must be specified along those edges. Shells for which boundary conditions must be specified along full circular edges are not investigated in the present chapter; that will be done in Chap. 12 on chimneys and Chap. 13 on storage tanks. As announced in Chap. 6, for that type of problems, we first have to upgrade the shell theory in Chap. 9. For circular cylindrical roof problems, the less involved theory of Chap. 6 is sufficiently accurate.

It is convenient to introduce a polar co-ordinate system to the cross-sectional profile. The shell is symmetric with respect to the lines $\theta = 0$ and $x = 0$. Once again, for the circumferential ordinate y, the equality $y = a\theta$ holds, in which the constant radius of the circular cylinder is denoted by a, and the constant curvature is equal to $k_y = -1/a$. Using this description of the middle surface, we can use the bending theory of Chap. 6 to analyse the bending behaviour of the circular

J. Blaauwendraad and J. H. Hoefakker, *Structural Shell Analysis*,
Solid Mechanics and Its Applications 200, DOI: 10.1007/978-94-007-6701-0_7,
© Springer Science+Business Media Dordrecht 2014

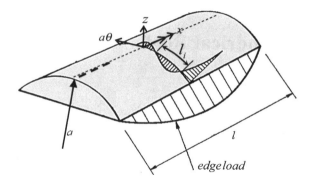

Fig. 7.1 Edge disturbance due to edge load

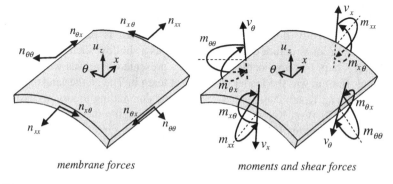

membrane forces   *moments and shear forces*

Fig. 7.2 Forces and moments on a *cylindrical* shell element

cylindrical shell. We review the relations derived in Chap. 6 with respect to the
change to the polar co-ordinate system. Figure 7.2 shows the membrane forces,
moments and shear forces which play a role in a cylindrical shell. The following
vectors are used:

$$
\begin{aligned}
\mathbf{u} &= \begin{bmatrix} u_x & u_\theta & u_z \end{bmatrix}^{\mathrm{T}} \\
\mathbf{e} &= \begin{bmatrix} \varepsilon_{xx} & \varepsilon_{\theta\theta} & \gamma_{x\theta} & \kappa_{xx} & \kappa_{\theta\theta} & \rho_{x\theta} \end{bmatrix}^{\mathrm{T}} \\
\mathbf{s} &= \begin{bmatrix} n_{xx} & n_{\theta\theta} & n_{x\theta} & m_{xx} & m_{\theta\theta} & m_{x\theta} \end{bmatrix}^{\mathrm{T}} \\
\mathbf{p} &= \begin{bmatrix} p_x & p_\theta & p_z \end{bmatrix}^{\mathrm{T}}
\end{aligned}
\tag{7.1}
$$

7.2 Differential Equation for Circular Cylinder

In Chap. 6, we introduced in Eq. (6.19) the shell differential operator Γ and the
Laplacian Δ. For a circular cylindrical shell with the co-ordinate system placed on
the middle surface, as shown in Fig. 7.2, the centre of curvature lies on the

negative part of the normal to the middle surface. With $k_x = 0$, $k_y = -1/a$ and $k_{xy} = 0$ the operators Γ and Δ become

$$\Gamma = -\frac{1}{a}\frac{\partial^2}{\partial x^2}$$

$$\Delta = \frac{\partial^2}{\partial x^2} + \frac{1}{a^2}\frac{\partial^2}{\partial \theta^2}$$

(7.2)

Using the operators Eq. (7.2) in Eq. (6.22) of the bending theory, the differential equation for the normal displacement u_z, with only the normal load p_z, becomes

$$D_b\left(\frac{\partial^2}{\partial x^2} + \frac{1}{a^2}\frac{\partial^2}{\partial \theta^2}\right)^4 u_z + D_m(1-v^2)\frac{1}{a^2}\frac{\partial^4 u_z}{\partial x^4} = \left(\frac{\partial^2}{\partial x^2} + \frac{1}{a^2}\frac{\partial^2}{\partial \theta^2}\right)^2 p_z \quad (7.3)$$

The solution of this differential equation consists of two parts: the inhomogeneous and the homogenous solution. For the inhomogeneous solution we can use the membrane solution. In general, this solution will not satisfy the boundary conditions along the straight edges. Therefore, the homogeneous solution must be added, such that the boundary conditions are satisfied. It means, that we must investigate the solution of the homogenous differential equation for loads at the edge.

7.3 Boundary Conditions at a Straight Edge

Equation (7.3) is a linear differential equation of the eighth order; therefore the homogeneous solution of this equation will be a linear combination of eight functions. The constants in the combination have to be determined by the boundary conditions along the straight edge $\theta = constant$. Four boundary conditions have to be set up per edge. These boundary conditions can involve four displacements u_x, u_θ, u_z, φ_x and four shell forces n_{xx}, $n_{\theta x}$, v_θ^k, $m_{\theta\theta}$. The rotation φ_θ is defined by

$$\varphi_\theta = -\frac{\partial u_z}{a\partial \theta} \quad (7.4)$$

The transverse shear force $^*v_\theta$ is the *Kirchhoff shear force*. As is known from thin plate bending theory, Kirchhoff showed that we obtain the correct number of boundary conditions only if we combine the transverse shear force v_θ and the twisting moment $m_{\theta x}$ to one quantity at the edge:

$$^*v_\theta = v_\theta + \frac{\partial m_{\theta x}}{\partial x} \quad (7.5)$$

Using these expressions, we can satisfy four boundary conditions per edge and therefore can find four unknown constants.

7.4 Expressions for Shell Forces and Displacements

Setting all load components equal to zero, we find that the homogeneous differential equation (7.3) is

$$D_b\left(\frac{\partial^2}{\partial x^2}+\frac{1}{a^2}\frac{\partial^2}{\partial\theta^2}\right)^4 u_z+\frac{D_m(1-v^2)}{a^2}\frac{\partial^4 u_z}{\partial x^4}=0 \qquad (7.6)$$

After we have solved this differential equation, we can determine the shell forces and other displacements from the obtained solution for the displacement $u_z(x,\theta)$. The bending curvatures and the torsional deformation are directly determined by using the kinematic relation (6.5). Substituting these results into the constitutive relation (6.6) for the bending and twisting moments, we find

$$m_{xx}=-D_b\left(\frac{\partial^2 u_z}{\partial x^2}+\frac{v}{a^2}\frac{\partial^2 u_z}{\partial\theta^2}\right)$$

$$m_{\theta\theta}=-D_b\left(v\frac{\partial^2 u_z}{\partial x^2}+\frac{1}{a^2}\frac{\partial^2 u_z}{\partial\theta^2}\right) \qquad (7.7)$$

$$m_{x\theta}=-D_b\frac{(1-v)}{a}\frac{\partial^2 u_z}{\partial x\partial\theta}$$

By substituting these expressions into the equilibrium equations (6.8) we obtain the expressions for the transverse shear forces:

$$v_\theta=\frac{\partial m_{x\theta}}{\partial x}+\frac{1}{a}\frac{\partial m_{\theta\theta}}{\partial\theta}=-D_b\left(\frac{1}{a}\frac{\partial^3 u_z}{\partial x^2\partial\theta}+\frac{1}{a^3}\frac{\partial^3 u_z}{\partial\theta^3}\right)$$

$$v_x=\frac{\partial m_{xx}}{\partial x}+\frac{1}{a}\frac{\partial m_{x\theta}}{\partial\theta}=-D_b\left(\frac{\partial^3 u_z}{\partial x^3}+\frac{1}{a^2}\frac{\partial^3 u_z}{\partial x\partial\theta^2}\right) \qquad (7.8)$$

Substitution of the expression for v_θ^k into the Kirchhoff shear force (7.5) yields

$$^*v_\theta=v_\theta+\frac{\partial m_{x\theta}}{\partial x}=-D_b\left(\frac{(2-v)}{a}\frac{\partial^3 u_z}{\partial x^2\partial\theta}+\frac{1}{a^3}\frac{\partial^3 u_z}{\partial\theta^3}\right) \qquad (7.9)$$

Substitution of Eq. (7.8) into Eq. (6.8) for the normal equilibrium yields the stress resultant $n_{\theta\theta}$ in the circumferential direction. Of course, the load component p_z is set to zero and therefore the substitution yields

$$n_{\theta\theta}=a\frac{\partial v_x}{\partial x}+\frac{\partial v_\theta}{\partial\theta}=-D_b a\left(\frac{\partial^4 u_z}{\partial x^4}+\frac{2}{a^2}\frac{\partial^4 u_z}{\partial x^2\partial\theta^2}+\frac{1}{a^4}\frac{\partial^4 u_z}{\partial\theta^4}\right) \qquad (7.10)$$

Using the Laplacian in Eq. (7.2), this relation is rewritten as

$$n_{\theta\theta}=a\frac{\partial v_x}{\partial x}+\frac{\partial v_\theta}{\partial\theta}=-D_b a\Delta\Delta u_z \qquad (7.11)$$

Substituting this expression in the other two equilibrium equations (6.8), we find the shear membrane force $n_{x\theta}$ and the normal membrane force n_{xx}:

$$n_{x\theta} = \int -\frac{1}{a}\frac{\partial n_{\theta\theta}}{\partial\theta}\,dx = D_b \int \frac{\partial\Delta\Delta u_z}{\partial\theta}\,dx$$

$$n_{xx} = \int -\frac{1}{a}\frac{\partial n_{x\theta}}{\partial\theta}\,dx = -D_b\frac{1}{a}\int\int \frac{\partial^2\Delta\Delta u_z}{\partial\theta^2}\,dx^2 \qquad (7.12)$$

Rewriting the constitutive relation (6.12) for the normal strains, we find

$$\varepsilon_{xx} = \frac{1}{D_m(1-v^2)}(n_{xx} - vn_{\theta\theta})$$

$$\varepsilon_{\theta\theta} = \frac{1}{D_m(1-v^2)}(-vn_{xx} + n_{\theta\theta}) \qquad (7.13)$$

Substituting this constitutive relation for the normal strains in the kinematic relation (4.2), we can determine the displacements u_x and u_θ:

$$u_x = \int \varepsilon_{xx}\,dx = \frac{1}{D_m(1-v^2)}\int (n_{xx} - vn_{\theta\theta})\,dx$$

$$u_\theta = \int (a\varepsilon_{\theta\theta} - u_z)\,d\theta = \int \left(\frac{a}{D_m(1-v^2)}(n_{\theta\theta} - vn_{xx}) - u_z\right)d\theta \qquad (7.14)$$

Since the normal displacement u_z is known, the displacements are determined by substituting the homogeneous solution for the in-plane stress resultants (7.11) and (7.12) in these expressions:

$$u_x = -\frac{t^2}{12(1-v^2)}\left(\frac{1}{a}\int\int\int \frac{\partial^2\Delta\Delta u_z}{\partial\theta^2}\,dx^3 - va\int \Delta\Delta u_z\,dx\right)$$

$$u_\theta = -\frac{t^2}{12(1-v^2)}\left(a^2\int \Delta\Delta u_z\,d\theta - v\int\int \frac{\partial\Delta\Delta u_z}{\partial\theta}\,dx^2\right) - \int u_z\,d\theta \qquad (7.15)$$

7.5 Homogeneous Solution for a Straight Edge

7.5.1 Exact Solution

When all the load components are zero, the differential Eq. (7.3) is

$$D_b\left(\frac{\partial^2}{\partial x^2} + \frac{1}{a^2}\frac{\partial^2}{\partial\theta^2}\right)^4 u_z + \frac{D_m(1-v^2)}{a^2}\frac{\partial^4 u_z}{\partial x^4} = 0 \qquad (7.16)$$

For the straight edge, we will consider loads or displacements that are developed in series of the type

$$f(x) = \sum_n f_n \cos\left(n\pi\frac{x}{l}\right) \tag{7.17}$$

where the origin of the x-axis is at mid-span. The number n can be 1, 3, 5, etc. Then, transverse edge forces and edge torques are maximum at mid-span and zero at the ends (diaphragms). It is useful to introduce the parameters β and α_n:

$$\beta^4 = \frac{D_m(1 - v^2)}{4D_b a^2} = \frac{3(1 - v^2)}{(at)^2}; \; \alpha_n = \frac{n\pi}{l} \tag{7.18}$$

The reduced equation becomes

$$\left(\frac{\partial^2}{\partial x^2} + \frac{1}{a^2}\frac{\partial^2}{\partial\theta^2}\right)^4 u_z + 4\beta^4\frac{\partial^4 u_z}{\partial x^4} = 0 \tag{7.19}$$

Hereafter we refer to it as *Donnell equation*. The trial solution for u_z is

$$u_z(x, \theta) = \sum_n A_n e^{r\theta} \cos(\alpha_n x) \tag{7.20}$$

This solution satisfies the boundary condition at the ends of the shell (diaphragm) that $u_z = 0$. On the basis of Eq. (7.7) also the bending moment m_{xx} is zero, which is expected in a pin-connection. Substitution of Eq. (7.20) in Eq. (7.19) yields

$$\left(\left(\frac{r}{a}\right)^2 - \alpha_n^2\right)^4 + 4\beta^4\alpha_n^4 = 0 \tag{7.21}$$

The following parameters are introduced:

$$\gamma = \frac{\alpha_n}{\beta} = \frac{n\pi}{\sqrt[4]{3(1 - v^2)}}\frac{\sqrt{at}}{l}$$

$$\lambda = \sqrt{\beta\alpha_n} = \sqrt[8]{3(1 - v^2)}\frac{\sqrt{n\pi}}{\sqrt[4]{atl^2}} \tag{7.22}$$

$$\rho = \frac{r}{a\lambda}$$

Note that γ is a dimensionless parameter, in which we recognize the length $\sqrt{at}$ which we already saw playing a role in the edge disturbance at the circular end of the shell. The parameter γ relates the length $\sqrt{at}$ to the span l of the shell roof. In practice it always holds that $\gamma < 1$. The reciprocal of parameter λ is a characteristic length l_c in which the length $\sqrt[4]{atl^2}$ occurs. It will appear that, like $\sqrt{at}$ in axial direction, $\sqrt[4]{atl^2}$ is important for disturbances in circumferential direction. Accounting for the new parameters, the characteristic equation and its roots become

$$\left(\left(\frac{r}{a\beta}\right)^2 - \frac{\alpha_n}{\lambda}\right)^4 = -4 \Leftrightarrow \left(\rho^2 - \gamma\right)^4 = -4 \tag{7.23}$$

$$\textit{Roots}: \quad \rho^2 = \gamma \pm (1 \pm i)$$

The following parameters are introduced:

$$\sigma_1 = \sqrt{\frac{\sqrt{1 + (1 + \gamma)^2} + (1 + \gamma)}{2}}$$

$$\sigma_2 = \sqrt{\frac{\sqrt{1 + (1 - \gamma)^2} - (1 - \gamma)}{2}} \tag{7.24a}$$

The inverse relations between γ, σ_1 and σ_2 are

$$(1 + \gamma) = \sigma_1^2 - \frac{1}{4\sigma_1^2}$$

$$-(1 - \gamma) = \sigma_2^2 - \frac{1}{4\sigma_2^2} \tag{7.24b}$$

The solution of the characteristic equation (7.21) becomes

$$\rho_1^2 = \sigma_1^2 \pm i - \frac{1}{4\sigma_1^2} = \left(\sigma_1 \pm i\frac{1}{2\sigma_1}\right)^2$$

$$\rho_2^2 = \sigma_2^2 \pm i - \frac{1}{4\sigma_2^2} = \left(\sigma_2 \pm i\frac{1}{2\sigma_2}\right)^2 \tag{7.25}$$

The characteristic equation has eight roots:

$$\rho_{1,2,3,4} = \pm\left(\sigma_1 \pm i\frac{1}{2\sigma_1}\right)$$

$$\rho_{5,6,7,8} = \pm\left(\sigma_2 \pm i\frac{1}{2\sigma_2}\right) \tag{7.26}$$

These eight roots describe four pairs of conjugate complex functions. The sum and the difference of the functions of each pair are purely real or purely imaginary and constitute another set of eight independent homogeneous solutions. The solution of the reduced differential equation can thus be written as

$$
\begin{aligned}
u_z(x, \theta) = \Big\{ & e^{-\lambda \sigma_1 a\theta} \Big[B_{11} \cos\Big(\frac{\lambda}{2\sigma_1} a\theta\Big) + B_{12} \sin\Big(\frac{\lambda}{2\sigma_1} a\theta\Big) \Big] \\
& + e^{\lambda \sigma_1 a\theta} \Big[B_{13} \cos\Big(\frac{\lambda}{2\sigma_1} a\theta\Big) + B_{14} \sin\Big(\frac{\lambda}{2\sigma_1} a\theta\Big) \Big] \\
& + e^{-\lambda \sigma_2 a\theta} \Big[B_{21} \cos\Big(\frac{\lambda}{2\sigma_2} a\theta\Big) + B_{22} \sin\Big(\frac{\lambda}{2\sigma_2} a\theta\Big) \Big] \\
& + e^{\lambda \sigma_2 a\theta} \Big[B_{23} \cos\Big(\frac{\lambda}{2\sigma_2} a\theta\Big) + B_{24} \sin\Big(\frac{\lambda}{2\sigma_2} a\theta\Big) \Big] \Big\} \cos(\alpha_n x)
\end{aligned}
\tag{7.27}
$$

This solution for the displacement $u_z(x, \theta)$ describes the disturbance for the straight edge of a circular cylindrical shell under loads or displacements having a cosine distribution in the axial direction. It is important to note that, like the parameters λ and σ, the constants B_{11} up to and including B_{24} depend on the number of circumferential whole waves n.

7.5.2 Approximate Solution

The parameter γ is defined in Eq. (7.22). For the static behaviour of thin shells under the usual loading cases, only the first few values of the wave number n are important; this means that this parameter remains small compared to unity since the thickness-to-radius ratio is small for thin shells. Thus for the static behaviour, rewriting Eq. (7.22), the following inequality holds:

$$
\gamma = \frac{n\pi}{\sqrt[4]{3(1 - v^2)}} \frac{a}{l} \sqrt{\frac{t}{a}} < 1
\tag{7.28}
$$

The parameters σ_1 and σ_2 that are defined by (7.24a) can be approximated. The roots are expanded according to the Taylor series: $(1 + x)^a = 1 + ax + \frac{1}{2} a(a - 1)x^2 + O(x^3)$. By stopping after the second term, since $\gamma^2 \ll 1$, we find that the parameters are

$$
\sigma_1 \approx \sqrt{\frac{\sqrt{2} + 1}{2}} \Big(1 + \frac{\gamma}{4}\sqrt{2}\Big)
$$
$$
\sigma_2 \approx \sqrt{\frac{\sqrt{2} - 1}{2}} \Big(1 + \frac{\gamma}{4}\sqrt{2}\Big)
\tag{7.29}
$$

In solution (7.27) two different powers are used:

$$
k_1 = \lambda \sigma_1 a\theta, \quad k_2 = \lambda \sigma_2 a\theta
\tag{7.30}
$$

The terms multiplied by $e^{\pm k_1}$ have an influence length $l_{i,1}$ (note that $\gamma < 1$)

$$l_{i,1} = \frac{\pi}{\lambda \sigma_1} \approx \frac{\pi}{\sqrt[8]{3(1-v^2)}\sqrt{n\pi}} \sqrt{\frac{2}{\sqrt{2}+1}} \sqrt[4]{atl^2} \approx 1,41 \frac{1}{\sqrt{n}} \sqrt[4]{atl^2} \qquad (7.31)$$

Or, written alternatively,

$$l_{i,1} \approx 1,41 \frac{a}{\sqrt{n}} \sqrt[4]{\frac{t}{a}\left(\frac{l}{a}\right)^2} \qquad (7.32)$$

The terms multiplied by $e^{\pm k_2}$ have an influence length $l_{i,2}$:

$$l_{i,2} = \frac{\pi}{\lambda \sigma_2} \approx \frac{\pi}{\sqrt[8]{3(1-v^2)}\sqrt{n\pi}} \sqrt{\frac{2}{\sqrt{2}-1}} \sqrt[4]{atl^2} \approx 3.39 \frac{1}{\sqrt{n}} \sqrt[4]{atl^2} \qquad (7.33)$$

$$l_{i,2} \approx 3.39 \frac{a}{\sqrt{n}} \sqrt[4]{\frac{t}{a}\left(\frac{l}{a}\right)^2} \qquad (7.34)$$

For the usual cylindrical roof, the thickness-to-radius ratio t/a varies between $1/100 \sim 1/200$ and the length-to-radius ratio l/a is about 1. Then the latter influence length, which is more than two times larger than the other, is approximately $l_{i,2} \approx a/\sqrt{n}$. This means that, if the distance between the straight edges of a circular cylindrical shell is larger than approximately the radius of the cylinder, the edge disturbance starting at one edge is not influenced by the edge disturbance starting at the other edge.

7.6 Displacements and Shell Forces of the Homogeneous Solution

The solution (7.27) for the normal displacement $u_z(x, \theta)$ can be differentiated and integrated. To obtain the shell forces and other displacements, the expressions (7.7)–(7.15) are used and the results are tabulated for $v = 0$ in Table 7.1. To simplify the formulations, the following expressions have been used:

$$\begin{array}{ll}
P_1 = \sigma_1^3 - 2\sigma_1 + \frac{1}{4\sigma_1} & P_2 = \sigma_2^3 + 2\sigma_2 + \frac{1}{4\sigma_2} \\
Q_1 = \frac{1}{2}\sigma_1 + \frac{1}{\sigma_1} + \frac{1}{8\sigma_1^3} & Q_2 = \frac{1}{2}\sigma_2 - \frac{1}{\sigma_2} + \frac{1}{8\sigma_2^3} \\
k_1 = \lambda \sigma_1 a\theta & k_2 = \lambda \sigma_2 a\theta \\
c_1 = \cos\left(\frac{\lambda}{2\sigma_1} a\theta\right) & c_2 = \cos\left(\frac{\lambda}{2\sigma_2} a\theta\right) \\
s_1 = \sin\left(\frac{\lambda}{2\sigma_1} a\theta\right) & s_2 = \sin\left(\frac{\lambda}{2\sigma_2} a\theta\right)
\end{array} \qquad (7.35)$$

To help designers, Table 7.1 has been used to analyze the shell roof for two different edge loads. The first case is a cosine load $n_{\theta\theta}$, and the second is a sine

Table 7.1 The expressions for the quantities of the homogeneous solution for the straight edge

	Multiplicator	$e^{-k_1} \cdot c_1$	$e^{-k_1} \cdot s_1$	$e^{k_1} \cdot c_1$	$e^{k_1} \cdot s_1$
u_z	$\cos(\alpha_n x)$	B_{11}	B_{12}	B_{13}	B_{14}
φ_x	$\lambda\cos(\alpha_n x)$	$\sigma_1 B_{11} - \frac{1}{2\sigma_1}B_{12}$	$\frac{1}{2\sigma_1}B_{11} + \sigma_1 B_{12}$	$-\sigma_1 B_{13} - \frac{1}{2\sigma_1}B_{14}$	$\frac{1}{2\sigma_1}B_{13} - \sigma_1 B_{14}$
u_x	$\frac{\alpha_n}{2a\lambda^2}\sin(\alpha_n x)$	$-B_{11} - (1+\gamma)B_{12}$	$(1+\gamma)B_{11} - B_{12}$	$-B_{13} + (1+\gamma)B_{14}$	$-(1+\gamma)B_{13} - B_{14}$
u_θ	$\frac{1}{2a\lambda}\cos(\alpha_n x)$	$Q_1 B_{11} - P_1 B_{12}$	$P_1 B_{11} + Q_1 B_{12}$	$-Q_1 B_{13} - P_1 B_{14}$	$P_1 B_{13} - Q_1 B_{14}$
$m_{\theta\theta}$	$D_b\lambda^2\cos(\alpha_n x)$	$-(1+\gamma)B_{11} + B_{12}$	$-B_{11} - (1+\gamma)B_{12}$	$-(1+\gamma)B_{13} - B_{14}$	$B_{13} - (1+\gamma)B_{14}$
v_θ^k	$D_b\lambda^3\cos(\alpha_n x)$	$-P_1 B_{11} - Q_1 B_{12}$	$Q_1 B_{11} - P_1 B_{12}$	$P_1 B_{13} - Q_1 B_{14}$	$Q_1 B_{13} + P_1 B_{14}$
n_{xx}	$\frac{E t\,\alpha_n^2}{2a\lambda^2}\cos(\alpha_n x)$	$-B_{11} - (1+\gamma)B_{12}$	$(1+\gamma)B_{11} - B_{12}$	$-B_{13} + (1+\gamma)B_{14}$	$-(1+\gamma)B_{13} - B_{14}$
$n_{\theta\theta}$	$\frac{E t\,\alpha_n^4}{2a\lambda^4}\cos(\alpha_n x)$	B_{12}	$-B_{11}$	$-B_{14}$	B_{13}
$n_{x\theta}$	$\frac{E t\,\alpha_n^3}{2a\lambda^3}\sin(\alpha_n x)$	$\frac{1}{2\sigma_1}B_{11} + \sigma_1 B_{12}$	$-\sigma_1 B_{11} + \frac{1}{2\sigma_1}B_{12}$	$-\frac{1}{2\sigma_1}B_{13} + \sigma_1 B_{14}$	$-\sigma_1 B_{13} - \frac{1}{2\sigma_1}B_{14}$

	Multiplicator	$e^{-k_2} \cdot c_2$	$e^{-k_2} \cdot s_2$	$e^{k_2} \cdot c_2$	$e^{k_2} \cdot s_2$
u_z	$\cos(\alpha_n x)$	B_{21}	B_{22}	B_{23}	B_{24}
φ_x	$\lambda\cos(\alpha_n x)$	$\sigma_2 B_{21} - \frac{1}{2\sigma_2}B_{22}$	$\frac{1}{2\sigma_2}B_{21} + \sigma_2 B_{22}$	$-\sigma_2 B_{23} - \frac{1}{2\sigma_2}B_{24}$	$\frac{1}{2\sigma_2}B_{23} - \sigma_2 B_{24}$
u_x	$\frac{\alpha_n}{2a\lambda^2}\sin(\alpha_n x)$	$B_{21} - (1-\gamma)B_{22}$	$(1-\gamma)B_{12} + B_{22}$	$B_{23} + (1-\gamma)B_{24}$	$-(1-\gamma)B_{23} + B_{24}$
u_θ	$\frac{1}{2a\lambda}\cos(\alpha_n x)$	$-Q_2 B_{21} + P_2 B_{22}$	$-P_2 B_{21} - Q_2 B_{22}$	$Q_2 B_{23} + P_2 B_{24}$	$-P_2 B_{23} + Q_2 B_{24}$
$m_{\theta\theta}$	$D_b\lambda^2\cos(\alpha_n x)$	$(1-\gamma)B_{21} + B_{22}$	$-B_{21} + (1-\gamma)B_{22}$	$(1-\gamma)B_{23} - B_{24}$	$B_{13} + (1-\gamma)B_{14}$
v_θ^k	$D_b\lambda^3\cos(\alpha_n x)$	$-P_2 B_{21} - Q_2 B_{22}$	$Q_2 B_{21} - P_2 B_{22}$	$P_2 B_{23} - Q_2 B_{24}$	$Q_2 B_{23} + P_2 B_{24}$
n_{xx}	$\frac{E t\,\alpha_n^2}{2a\lambda^2}\cos(\alpha_n x)$	$B_{21} - (1-\gamma)B_{22}$	$(1-\gamma)B_{12} + B_{22}$	$B_{23} + (1-\gamma)B_{24}$	$-(1-\gamma)B_{23} + B_{24}$
$n_{\theta\theta}$	$\frac{E t\,\alpha_n^4}{2a\lambda^4}\cos(\alpha_n x)$	$-B_{22}$	B_{21}	B_{24}	$-B_{23}$
$n_{x\theta}$	$\frac{E t\,\alpha_n^3}{2a\lambda^3}\sin(\alpha_n x)$	$-\frac{1}{2\sigma_2}B_{21} - \sigma_2 B_{22}$	$\sigma_2 B_{21} - \frac{1}{2\sigma_2}B_{22}$	$\frac{1}{2\sigma_2}B_{23} - \sigma_2 B_{24}$	$+\sigma_2 B_{23} + \frac{1}{2\sigma_2}B_{24}$

load $n_{\theta x}$. Tables 7.2 and 7.3 contain the results of the analyses. The two cases are useful, because the membrane solution leaves the straight edges with non-zero forces $n_{\theta\theta}$ and $n_{\theta x}$, so a designer can add the solutions of the tables to make the edges stress-free. The following conditions are used:

$$
\begin{aligned}
Case\ 1: \quad & n_{\theta\theta} \neq 0; \quad n_{x\theta} = m_{\theta\theta} = v_\theta^k = 0 \\
Case\ 2: \quad & n_{x\theta} \neq 0; \quad n_{\theta\theta} = m_{\theta\theta} = v_x^k = 0
\end{aligned}
\tag{7.36}
$$

The results for $n_{\theta\theta} \neq 0$ are tabulated in Table 7.2 and the results for $n_{x\theta} \neq 0$ are tabulated in Table 7.3. These tables are derived for two edge loads, which are f_θ and f_x respectively, that are described by half the wavelength of the trigonometric functions. This means that the wave number n is equal to 1 and $\alpha_n x = \pi x/l$. In the tables, the edge loads have a top value along the edge that is equal to the value of the multiplier. The tables represent the edge disturbance starting at the straight edge at $y = a\theta = 0$ for simple edge conditions that respectively are equal to:

$$Table\ 7.2: \quad f_\theta = \frac{Et}{2a}\left(\frac{\alpha_n}{\lambda}\right)^4 (2.34 + 8.49\gamma)\cos(\alpha_n x)$$

$$Table\ 7.3: \quad f_x = \frac{Et}{2a}\left(\frac{\alpha_n}{\lambda}\right)^3 (2.34 + 8.49\gamma)\sin(\alpha_n x)$$

(7.37)

Using the tables, we can calculate the stresses and displacements due to an edge disturbance starting at straight edge ($\theta = 0$). This is exemplified in Sect. 7.7.

7.7 Application to a Shell Roof Under its Own Weight

The geometry of the circular cylindrical shell roof shown in Fig. 7.3 is defined by the radius a, the thickness t and the length of the straight edge l. At the curved edges, the shell is supported by diaphragms. This means that these edges are hinge supported and that the straight edges are free. Figure 7.3 depicts two co-ordinate systems that are indicated with the superscript m for the membrane analysis and b for the bending analysis. This is done because the determination of the response of the shell roof to its own weight is performed in two consecutive steps. The first step is the determination of the membrane response by using boundary conditions for the curved edges. This will yield non-zero stress resultants at the free edge, which must be counteracted by bending stress resultants. For the second step, Tables 7.2 and 7.3 are used to determine the shell forces due to the bending behaviour. For the membrane analysis, it is convenient to use a co-ordinate system that is placed at the crown of the roof. On the other hand, for the bending analysis, it is necessary to use a co-ordinate system that is placed at the edges, from which the edge disturbance originates.

The own weight of this circular roof can be modelled in two ways; perpendicular to the curved surface and as a vertical load. A simplified analysis is performed by assuming that the load is acting perpendicular to the curved surface, which is in fact an axisymmetric load or uniform load. In this case, it is assumed that it is not necessary to include the boundary conditions at the curved edges. Under this assumption, only one membrane stress resultant is activated: the stress resultant $n_{\theta\theta}$ in circumferential direction. This simplification is allowed if the shell is shallow. For a less shallow shell, the vertical load has to be rewritten as is depicted in Fig. 4.4. Corresponding to the membrane analysis of closed circular cylindrical shells, this description of the load yields a solution for all membrane stress resultants ($n_{\theta\theta}$, $n_{x\theta}$, n_{xx}). In Sects. 7.7.1, 7.7.2, the response of a very shallow circular shell roof to a uniform load and the response of a more curved circular shell roof to a vertical load will be determined respectively.

Table 7.2 The expressions for the quantities for the straight edge ($\theta = 0$) under simple edge load $n_{00} \neq 0$

$$\left. \begin{array}{l} n = 1 \\ v = 0 \end{array} \right\} \Rightarrow \quad \begin{cases} \alpha_n = \dfrac{\pi}{l} = \dfrac{3.14}{l} \\[2mm] \beta = \dfrac{\sqrt{3}}{\sqrt{at}} = \dfrac{1.32}{\sqrt{at}} \\[2mm] \lambda^2 = \alpha_n\beta = \dfrac{4.13}{l\sqrt{at}} \\[2mm] \hat{f}_\theta = \dfrac{Et}{2a}\left(\dfrac{\alpha_n}{\lambda}\right)^4 (2.34 + 8.49\gamma) \end{cases}$$

$$\sigma_1 = 1.10 + 0.39\gamma$$
$$\sigma_2 = 0.46 + 0.16\gamma$$
$$\gamma = \left(\dfrac{\alpha_n}{\lambda}\right)^2 = 2.39\,\dfrac{\sqrt{at}}{l}$$
$$at\ (x,\theta) = (0,0)$$

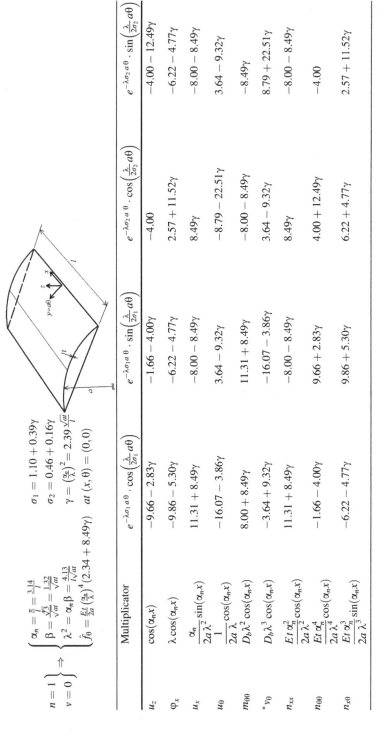

	Multiplicator	$e^{-\lambda\sigma_1 a\theta}\cdot\cos\left(\frac{\lambda}{2\sigma_1}a\theta\right)$	$e^{-\lambda\sigma_1 a\theta}\cdot\sin\left(\frac{\lambda}{2\sigma_1}a\theta\right)$	$e^{-\lambda\sigma_2 a\theta}\cdot\cos\left(\frac{\lambda}{2\sigma_2}a\theta\right)$	$e^{-\lambda\sigma_2 a\theta}\cdot\sin\left(\frac{\lambda}{2\sigma_2}a\theta\right)$
u_z	$\cos(\alpha_n x)$	$-9.66 - 2.83\gamma$	$-1.66 - 4.00\gamma$	-4.00	$-4.00 - 12.49\gamma$
φ_x	$\lambda\cos(\alpha_n x)$	$-9.86 - 5.30\gamma$	$-6.22 - 4.77\gamma$	$2.57 + 11.52\gamma$	$-6.22 - 4.77\gamma$
u_x	$\dfrac{\alpha_n}{2a\lambda^2}\sin(\alpha_n x)$	$11.31 + 8.49\gamma$	$-8.00 - 8.49\gamma$	8.49γ	$-8.00 - 8.49\gamma$
u_θ	$\dfrac{1}{2a}\lambda\cos(\alpha_n x)$	$-16.07 - 3.86\gamma$	$3.64 - 9.32\gamma$	$-8.79 - 22.51\gamma$	$3.64 - 9.32\gamma$
m_{00}	$D_b\lambda^2\cos(\alpha_n x)$	$8.00 + 8.49\gamma$	$11.31 + 8.49\gamma$	$-8.00 - 8.49\gamma$	-8.49γ
$^*v_\theta$	$D_b\lambda^3\cos(\alpha_n x)$	$-3.64 + 9.32\gamma$	$-16.07 - 3.86\gamma$	$3.64 - 9.32\gamma$	$8.79 + 22.51\gamma$
n_{xx}	$\dfrac{Et\,\alpha_n^2}{2a\lambda^2}\cos(\alpha_n x)$	$11.31 + 8.49\gamma$	$-8.00 - 8.49\gamma$	8.49γ	$-8.00 - 8.49\gamma$
n_{00}	$\dfrac{Et\,\alpha_n^4}{2a\lambda^4}\cos(\alpha_n x)$	$-1.66 - 4.00\gamma$	$9.66 + 2.83\gamma$	$4.00 + 12.49\gamma$	-4.00
$n_{x\theta}$	$\dfrac{Et\,\alpha_n^3}{2a\lambda^3}\sin(\alpha_n x)$	$-6.22 - 4.77\gamma$	$9.86 + 5.30\gamma$	$6.22 + 4.77\gamma$	$2.57 + 11.52\gamma$

Table 7.3 The expressions for the quantities for the straight edge ($\theta = 0$) under simple edge load $n_{x\theta} \neq 0$

$$n=1 \atop v=0 \Rightarrow \begin{cases} \alpha_n = \frac{\pi}{l} = \frac{3.14}{l} & \sigma_1 = 1.10 + 0.39\gamma \\ \beta = \frac{\sqrt{3}}{\sqrt{at}} = \frac{1.32}{\sqrt{at}} & \sigma_2 = 0.46 + 0.16\gamma \\ \lambda^2 = \alpha_n\beta = \frac{4.13}{l\sqrt{at}} & \gamma = \left(\frac{\alpha_n}{\lambda}\right)^2 = 2.39\frac{\sqrt{at}}{l} \\ \hat{f}_x = \frac{Et}{2a}\left(\frac{\alpha_n}{\lambda}\right)^3(2.34 + 8.49\gamma) & at\ (x,\theta) = (0,0) \end{cases}$$

	Multiplicator	$e^{-\lambda\sigma_1 a\theta}\cdot\cos\left(\frac{\lambda}{2\sigma_1}a\theta\right)$	$e^{-\lambda\sigma_1 a\theta}\cdot\sin\left(\frac{\lambda}{2\sigma_1}a\theta\right)$	$e^{-\lambda\sigma_2 a\theta}\cdot\cos\left(\frac{\lambda}{2\sigma_2}a\theta\right)$	$e^{-\lambda\sigma_2 a\theta}\cdot\sin\left(\frac{\lambda}{2\sigma_2}a\theta\right)$
u_z	$\cos(\alpha_n x)$	$4.39 + 4.28\gamma$	$1.82 + 5.95\gamma$	$0.75 - 2.46\gamma$	$1.82 + 5.95\gamma$
φ_x	$\lambda\cos(\alpha_n x)$	$4.00 + 4.00\gamma$	$4.00 + 8.49\gamma$	$-1.66 - 6.83\gamma$	1.66
u_x	$\frac{\alpha_n}{2a\lambda^2}\sin(\alpha_n x)$	$-6.22 - 12.05\gamma$	$2.57 + 2.73\gamma$	$-1.07 - 6.59\gamma$	$2.57 + 2.73\gamma$
u_θ	$\frac{1}{2a\lambda}\cos(\alpha_n x)$	$8.00 + 8.49\gamma$	8.49γ	$3.31 + 8.49\gamma$	8.49γ
$m_{\theta\theta}$	$D_b\lambda^2\cos(\alpha_n x)$	$-2.57 - 2.73\gamma$	$-6.22 - 12.05\gamma$	$2.57 + 2.73\gamma$	$1.07 + 6.59\gamma$
$*v_\theta$	$D_b\lambda^3\cos(\alpha_n x)$	-8.49γ	$8.00 + 8.49\gamma$	8.49γ	$-3.31 - 8.49\gamma$
n_{xx}	$\frac{Et\,\alpha_n^2}{2a\lambda^2}\cos(\alpha_n x)$	$-6.22 - 12.05\gamma$	$2.57 + 2.73\gamma$	$-1.07 - 6.59\gamma$	$2.57 + 2.73\gamma$
$n_{\theta\theta}$	$\frac{Et\,\alpha_n^4}{2a\lambda^4}\cos(\alpha_n x)$	$1.82 + 5.95\gamma$	$-4.39 - 4.28\gamma$	$-1.82 - 5.95\gamma$	$0.75 - 2.46\gamma$
$n_{x\theta}$	$\frac{Et\,\alpha_n^3}{2a\lambda^3}\sin(\alpha_n x)$	$4.00 + 8.49\gamma$	$-4.00 - 4.00\gamma$	-1.66	$-1.66 - 6.83\gamma$

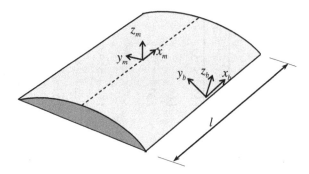

Fig. 7.3 Circular cylindrical shell roof with co-ordinate system (x^m, y^m, z^m) for the membrane analysis and (x^b, y^b, z^b) for the bending analysis

7.7.1 Uniform Load

The load components are

$$p_x = 0, \quad p_\theta = 0, \quad p_z = -p \tag{7.38}$$

From Eq. (4.8) we find

$$
\begin{aligned}
n_{\theta\theta} &= ap_z = -ap \\
n_{x\theta} &= n_{xx} = 0
\end{aligned}
\tag{7.39}
$$

Since the membrane solution for $n_{x\theta}$ is zero, we only need to use Table 7.2. The table has been derived for an edge load described by a trigonometric function of the form $f_\theta(x) = \hat{f}_\theta \cos(\alpha_n x)$. To make Table 7.2 applicable to the problem of the constant membrane edge load under consideration, this constant load is developed into a Fourier series. Using Fourier analysis, we find that a uniform load p over a length l is equivalently described by the series

$$p(x) = \frac{4p}{\pi} \sum_{n=1,3,5\ldots}^{\infty} \frac{1}{n} \cos\frac{n\pi x}{l} \tag{7.40}$$

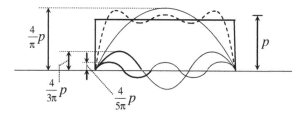

Fig. 7.4 The first, third and fifth harmonic of the series for a uniform load p and their sum (*dashed line*)

The first, third and fifth harmonic of this series are depicted in Fig. 7.4 and it can be shown that the stress resultants and displacements resulting from the series solution converge rapidly and that the first term gives a satisfactory accuracy. The force $n_{\theta\theta,m}$ of Eq. (7.39) is thus developed in a series

$$n_{\theta\theta,m} = -\frac{4pa}{\pi} \sum_{n=1,3,5...}^{\infty} \frac{1}{n}\cos\frac{n\pi x}{l} \quad at \ \theta = \theta_o \tag{7.41}$$

Table 7.2 is derived by taking into account only the first harmonic, thus $n = 1$ and $\alpha_n = \frac{\pi}{l}$. This means that $n_{\theta\theta,m}$ at the edge is approximately described by

$$n_{\theta\theta,m} = -\frac{4pa}{\pi}\cos\frac{\pi x}{l} \tag{7.42}$$

At the free edge with an opening angle $\theta = \theta_o$, the boundary conditions are

$$n_{\theta\theta(\theta=\theta_o)} = n_{\theta\theta,m} + n_{\theta\theta,b} = 0$$
$$n_{x\theta(\theta=\theta_o)} = n_{x\theta,m} + n_{x\theta,b} = 0 \tag{7.43}$$

Thus the forces generated by the bending disturbance, which have to counteract the value of the forces of the membrane solution at the straight edge, must have the following value at the edge:

$$n_{\theta\theta,b} = -n_{\theta\theta,m} = \frac{4pa}{\pi}\cos\frac{\pi x}{l}$$
$$n_{x\theta,b} = -n_{x\theta,m} = 0 \qquad at \ \theta = \theta_o \tag{7.44}$$

Table 7.2 thus represents the edge disturbance starting at the straight edge for a simple edge condition $f_\theta(x) = \hat{f}_\theta \cos(\alpha_n x)$:

$$f_\theta(x) = \hat{f}_\theta \cos\frac{\pi x}{l}; \quad \hat{f}_\theta = \frac{Et}{2a}\left(\frac{\alpha_n}{\lambda}\right)^4 (2.34 + 8.49\gamma) \tag{7.45}$$

To obtain the shell forces generated by the bending disturbance, Table 7.2 has to be multiplied by a multiplier M, which is found by using the relation $M \cdot \hat{f}_\theta = \hat{n}_{\theta\theta,m}$ between the top values of the edge load and the required bending shell forces at the edge. In this case, the top value $4pa/\pi$ from Eq. (7.44) is divided by $\hat{f}_\theta$ from Eq. (7.45) to find that the multiplier M is equal to

$$M = \frac{1}{\frac{Et}{2a}\left(\frac{\alpha_n}{\lambda}\right)^4(2.34 + 8.49\gamma)}\frac{4pa}{\pi} = \frac{4pa}{\pi}\cdot\frac{2a}{Et}\left(\frac{\lambda}{\alpha_n}\right)^4\frac{1}{2.34 + 8.49\gamma} \tag{7.46}$$

For example, the expressions for the force n_{xx} have to be multiplied by

$$M \cdot \hat{n}_{xx,b} = \frac{4pa}{\pi}\frac{2a}{Et}\left(\frac{\lambda}{\alpha_n}\right)^4\frac{1}{2.34 + 8.49\gamma}\cdot\frac{Et}{2a}\left(\frac{\alpha_n}{\lambda}\right)^2$$
$$= \frac{4pa}{\pi}\left(\frac{\lambda}{\alpha_n}\right)^2\frac{1}{2.34 + 8.49\gamma} = \frac{4\sqrt[4]{3}}{\pi^2}\frac{1}{2.34 + 8.49\gamma}pl\sqrt{\frac{a}{t}} \tag{7.47}$$

At the straight edge, the expression for n_{xx} from Table 7.2 with $y^b = a$ $\theta^b = 0$ is

$$
\begin{aligned}
n_{xx,b(y^b=0)} &= \frac{4\sqrt[4]{3}}{\pi^2} \frac{1}{2.34 + 8.49\gamma} pl\sqrt{\frac{a}{t}}[(11.31 + 8.49\gamma) + 8.49\gamma] \cos\frac{\pi x}{l} \\
&= \frac{4\sqrt[4]{3}}{\pi^2} \frac{11.31 + 16.98\gamma}{2.34 + 8.49\gamma} pl\sqrt{\frac{a}{t}}\cos\frac{\pi x}{l}
\end{aligned}
\tag{7.48}
$$

This is the total n_{xx} since the membrane analysis had $n_{xx,m} = 0$. The total stress σ_{xx} at this edge is thus described by

$$
\sigma_{xx(y^b=0)} = \frac{n_{xx(y^b=0)}}{t} = \frac{1}{\pi^2} 4\sqrt[4]{3} \frac{11.31 + 16.98\gamma}{2.34 + 8.49\gamma} p\frac{l}{t}\sqrt{\frac{a}{t}}\cos\frac{\pi x}{l}
\tag{7.49}
$$

7.7.2 Vertical Load

As shown in Fig. 4.4, it is possible to replace the vertical load by surface load components. These load components are expressed by:

$$
\begin{aligned}
p_\theta &= p \sin\theta \\
p_z &= -p \cos\theta
\end{aligned}
\tag{7.50}
$$

Using the Fourier series of Eq. (7.40) for only the first harmonic as performed in Sect. 7.7.1, we find that the loads are

$$
\begin{aligned}
p_\theta &= \frac{4p}{\pi} \sin\theta \cos\frac{\pi x}{l} \\
p_z &= -\frac{4p}{\pi} \cos\theta \cos\frac{\pi x}{l}
\end{aligned}
\tag{7.51}
$$

The boundary conditions for the roof are

$$
\begin{aligned}
n_{x\theta} &= 0 \quad \text{at} \quad x = 0 \\
n_{xx} &= 0 \quad \text{at} \quad x = \pm\frac{l}{2}
\end{aligned}
\tag{7.52}
$$

The equilibrium equations in Eq. (4.7) for the membrane behaviour and the boundary conditions subsequently yield the solution for the forces $n_{\theta\theta}$, $n_{x\theta}$ and n_{xx}:

$$n_{\theta\theta} = ap_z = -\frac{4pa}{\pi}\cos\theta\cos\frac{\pi x}{l}$$

$$\frac{\partial n_{x\theta}}{\partial x} = -p_\theta - \frac{\partial n_{\theta\theta}}{a\partial\theta} = -\frac{8p}{\pi}\sin\theta\cos\frac{\pi x}{l}$$

$$n_{x\theta} = -\frac{8pl}{\pi^2}\sin\theta\sin\frac{\pi x}{l} \tag{7.53}$$

$$\frac{\partial n_{xx}}{\partial x} = -p_x - \frac{\partial n_{x\theta}}{a\partial\theta} = \frac{8pl}{\pi^2 a}\cos\theta\sin\frac{\pi x}{l}$$

$$n_{xx} = -\frac{8pl^2}{\pi^3 a}\cos\theta\cos\frac{\pi x}{l}$$

Since the membrane solution for $n_{x\theta}$ is not zero, Table 7.3 is also needed to find the shell forces generated by the edge disturbance. For the force n_{xx}, there is no boundary condition at the free edge, and therefore this force does not activate an edge disturbance. At the free edge with an opening angle $\theta = \theta_o$, the boundary conditions are

$$n_{\theta\theta(\theta=\theta_o)} = n_{\theta\theta,m} + n_{\theta\theta,b} = 0$$
$$n_{x\theta(\theta=\theta_o)} = n_{x\theta,m} + n_{x\theta,b} = 0 \tag{7.54}$$

This implies that the forces generated by the bending disturbance, which have to counteract the membrane forces at the straight edge, must have edge values

$$n_{\theta\theta,b} = -n_{\theta\theta,m} = \frac{4pa}{\pi}\cos\theta_o\cos\frac{\pi x}{l}$$
$$n_{x\theta,b} = -n_{x\theta,m} = \frac{8pl}{\pi^2}\sin\theta_o\sin\frac{\pi x}{l} \quad at \ \theta = \theta_o \tag{7.55}$$

Table 7.2 thus represents the edge disturbance starting at the straight edge for a simple edge condition $f_\theta(x) = \hat{f}_\theta\cos(\alpha_n x)$:

$$f_\theta(x) = \hat{f}_\theta\cos\frac{\pi x}{l}; \quad \hat{f}_\theta = \frac{Et}{2a}\left(\frac{\alpha_n}{\lambda}\right)^4(2.34 + 8.49\gamma) \tag{7.56}$$

To obtain the forces generated by the bending disturbance, Table 7.2 has to be multiplied by a multiplier $M(n_{\theta\theta})$, which is found by using the relation $M(n_{\theta\theta}) \times \hat{f}_\theta = \hat{n}_{\theta\theta,m}$ between the top values of the edge load and the required bending stress resultant at the edge. In this case, the top value $\frac{4pa}{\pi}\cos\theta_o$ from Eq. (7.55) is divided by $\hat{f}_\theta$ from Eq. (7.56) to find that the multiplier $M(n_{\theta\theta})$ is

$$M(n_{\theta\theta}) = \frac{4pa}{\pi}\cdot\frac{2a}{Et}\left(\frac{\lambda}{\alpha_n}\right)^4\frac{1}{2.34 + 8.49\gamma}\cos\theta_o \tag{7.57}$$

For example, the expressions for n_{xx} in Table 7.2 have to be multiplied by

$$
\begin{aligned}
\mathrm{M}(n_{\theta\theta}) \cdot \hat{n}_{xx,b} &= \frac{4ap}{\pi}\frac{2a}{Et}\left(\frac{\lambda}{\alpha_n}\right)^4\frac{1}{2.34+8.49\gamma}\cos\theta_o \cdot \frac{Et}{2a}\left(\frac{\alpha_n}{\lambda}\right)^2 \\
&= \frac{4\sqrt[4]{3}}{\pi^2}\frac{1}{2.34+8.49\gamma}pl\sqrt{\frac{a}{t}}\cos\theta_o
\end{aligned}
\tag{7.58}
$$

At the straight edge, the expression for n_{xx} from Table 7.2 with $y^b = a\theta^b = 0$ becomes

$$
\begin{aligned}
n_{xx,b(y^b=0)}^{(n_{\theta\theta})} &= \frac{4\sqrt[4]{3}}{\pi^2}\frac{1}{2.34+8.49\gamma}pl\sqrt{\frac{a}{t}}\cos\theta_o[(11.31+8.49\gamma)+8.49\gamma]\cos\frac{\pi x}{l} \\
&= \frac{4\sqrt[4]{3}}{\pi^2}\frac{11.31+16.98\gamma}{2.34+8.49\gamma}pl\sqrt{\frac{a}{t}}\cos\theta_o\cos\frac{\pi x}{l}
\end{aligned}
\tag{7.59}
$$

Table 7.3 thus represents the edge disturbance starting at the straight edge for a simple edge condition $f_x(x) = \hat{f}_x \sin(\alpha_n x)$:

$$
f_{x\theta}(x) = \hat{f}_x\sin\frac{\pi x}{l}; \quad \hat{f}_x = \frac{Et}{2a}\left(\frac{\alpha_n}{\lambda}\right)^3(2.34+8.49\gamma)
\tag{7.60}
$$

To obtain the shell forces generated by the bending disturbance, Table 7.3 has to be multiplied by a multiplier $\mathrm{M}(n_{x\theta})$, which is found by using the relation $\mathrm{M}(n_{x\theta}) \times \hat{f}_x = \hat{n}_{x\theta(\theta=\theta_o)}$ between the top values of the edge load and the required bending forces at the edge. In this case, the top value $\frac{8pl}{\pi^2}\sin\theta_o$ from Eq. (7.55) is divided by $\hat{f}_x$ from Eq. (7.60) to find that the multiplier $\mathrm{M}(n_{x\theta})$ is equal to

$$
\mathrm{M}(n_{x\theta}) = \frac{8pl}{\pi^2}\cdot\frac{2a}{Et}\left(\frac{\lambda}{\alpha_n}\right)^3\frac{1}{2.34+8.49\gamma}\sin\theta_o
\tag{7.61}
$$

For example, the expressions for n_{xx} from Table 7.3 have to be multiplied by

$$
\begin{aligned}
\mathrm{M}(n_{x\theta}) \cdot \hat{n}_{xx,b} &= \frac{8pl}{\pi^2}\frac{2a}{Et}\left(\frac{\lambda}{\alpha_n}\right)^3\frac{1}{2.34+8.49\gamma}\sin\theta_o \cdot \frac{Et}{2a}\left(\frac{\alpha_n}{\lambda}\right)^2 \\
&= \frac{8pl}{\pi^2}\left(\frac{\lambda}{\alpha_n}\right)\frac{1}{2.34+8.49\gamma}\sin\theta_o
\end{aligned}
\tag{7.62}
$$

At the straight edge, the expression for n_{xx} from Table 7.3 with $y^b = a\theta^b = 0$ is

$$
\begin{aligned}
n_{xx,b(y^b=0)}^{(n_{x\theta})} &= \frac{8pl}{\pi^2}\left(\frac{\lambda}{\alpha_n}\right)\frac{1}{2.34+8.49\gamma}\sin\theta_o \cdot [(-6.22-12.05\gamma)+(-1.07-6.59\gamma)]\cos\frac{\pi x}{l} \\
&= -\frac{8}{\pi^2}\frac{7.29+18.64\gamma}{2.34+8.49\gamma}pl\left(\frac{\lambda}{\alpha_n}\right)\sin\theta_o\cos\frac{\pi x}{l}
\end{aligned}
\tag{7.63}
$$

From the membrane solution (7.53) for $\theta^m = \theta_o$, it is found that $n_{xx,m}$ is

$$n_{xx,m} = -\frac{8pl^2}{\pi^3 a} \cos \theta_o \cos \frac{\pi x}{l} \tag{7.64}$$

The total stress σ_{xx} at this edge is thus described by

$$\sigma_{xx(y^b=0)} = \frac{1}{t} \left[n^{(n_{\theta\theta})}_{xx,b(y^b=0)} + n^{(n_{x\theta})}_{xx,b(y^b=0)} + n_{xx,m} \right] \tag{7.65}$$

The stress due to the bending disturbances is

$$\begin{aligned}
\sigma_{xx,b(y^b=0)} = {} & \frac{1}{\pi^2} p \frac{l}{t} \cos \frac{\pi x}{l} \left[4\sqrt[4]{3} \, \frac{11.31 + 16.98\gamma}{2.34 + 8.49\gamma} \sqrt{\frac{a}{t}} \cos \theta_o \right. \\
& \left. -8 \frac{7.29 + 18.64\gamma}{2.34 + 8.49\gamma} \left(\frac{\lambda}{\alpha_n} \right) \sin \theta_o \right]
\end{aligned} \tag{7.66}$$

The stress due to the membrane response is

$$\sigma_{xx,m(y^b=0)} = \frac{1}{\pi^2} p \frac{l}{t} \cos \frac{\pi x}{l} \left[-\frac{8}{\pi} \frac{l}{a} \cos \theta_o \right] \tag{7.67}$$

7.7.3 Comparison of Solutions for a Concrete Roof

We will compare the solution for the three different load cases for the concrete roof shell. The vertical load p due to its own weight, with specific mass $\rho = 2400$ kg/m^3, thickness t $= 0.07$ m and gravitational acceleration $g = 10$ m/s^2 and roofing is

$$p = \rho g t + \text{roofing} = 1\,900 \text{ N/m}^2 \tag{7.68}$$

The geometry of the shell is defined by the length of the straight edge l_x, the radius a and the thickness t. The length l_x is identical to the length l used in the previous subsections, but the subscript is introduced to distinguish it from the distance between the straight edges l_y (see Fig. 7.5). For the following geometry, the parameter γ is, accounting for $l_x = 18$ m, $a = 11.6$ m and $t = 0.07$ m

$$\gamma = \frac{\pi}{\sqrt[4]{3}} \frac{\sqrt{at}}{l_x} = 0.120 \tag{7.69}$$

For the chord l_y, the opening angle θ_o can be determined and subsequently the height h between the crown and centre of the chord. The height h is determined as follows

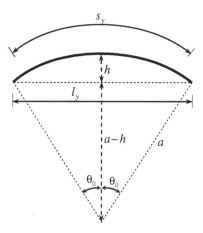

Fig. 7.5 Circumferential cross-section of the shell roof

$$l_y = 13.0 \text{ m} \quad \Rightarrow \quad \sin \theta_o = \frac{l_y}{2a} = 0.560$$
$$\Rightarrow \quad \theta_o = 0.595 \text{ rad} = 0.189 \cdot \pi \text{ rad}(\approx 34^o) \quad\quad\quad (7.70)$$
$$\Rightarrow \quad \cos \theta_o = 0.828 = \frac{a - h}{a}$$
$$\Rightarrow \quad h = (1 - \cos \theta_o)a = 0.172 \cdot a = 2.00 \text{ m}$$

The arc of the circle between the straight edges is $s_y = 2\theta_o \cdot a = 1.19 \cdot a$. The largest influence length of Eq. (7.34) with $n = 1$ and $v = 0$ is

$$l_{i,2} \approx 3.39 \cdot a \sqrt[4]{\frac{t}{a} \left(\frac{l_x}{a}\right)^2} = 1.18 \cdot a = 13.6 \text{ m} \quad\quad\quad (7.71)$$

The length of the shell in circumferential direction ($s_y = 1.19$ a) is, practically spoken, equal to the influence length of the edge disturbance in that direction ($l_{i2} = 1.18$ a). We conclude that the disturbance initiated at one straight edge does not influence the stress state at the opposite edge. Tables 7.2 and 7.3 can thus be used to derive the response of this shell.

Circular Shell Roof Under Uniform Load

In this case, treated in Sect. 7.7.1, the stress σ_{xx} is fully determined by the edge disturbance. For the edge $y^b = 0$, the expression Eq. (7.49) for σ_{xx} is derived and, substituting the load and the geometry given by Eqs. (7.68) and (7.69) respectively, the value of σ_{xx} at the middle of the free edge $\left(x^b = 0\right)$ becomes

$$\sigma_{xx(y^b=0)} = \frac{1}{\pi^2} 1900 \frac{18}{0.07} \cos 0 \cdot 4\sqrt[4]{3} \frac{11.31 + 16.98 \cdot 0.120}{2.34 + 8.49 \cdot 0.120} \sqrt{\frac{11.6}{0.07}} \tag{7.72}$$
$$= 13.3 \cdot 10^6 \, \text{N/m}^2 = 13.3 \, \text{N/mm}^2$$

Circular Shell Roof Under Vertical Load

In this case, treated in Sect. 7.7.2, the stress σ_{xx} is determined by both the membrane solution $(\sigma_{xx,m})$ and the edge disturbance $(\sigma_{xx,b})$. This edge disturbance is built up from two edge loads, one for $n_{\theta\theta}$ and one for $n_{x\theta}$. For the edge $y^b = 0$, the expression Eq. (7.66) for $\sigma_{xx,b}$ is derived and, with the load and the geometry given by Eqs. (7.68) and (7.69) respectively, the value of $\sigma_{xx,b}$ (at $x^b = 0$) becomes

$$
\begin{aligned}
\sigma_{xx,b(y^b=0)} &= \frac{1}{\pi^2} 1900 \cdot \frac{18}{0.07} \cos 0 \left[4\sqrt[4]{3} \frac{11.31 + 16.98 \cdot 0.120}{2.34 + 8.49 \cdot 0.120} \sqrt{\frac{11.6}{0.07}} \cdot 0.828 \right.\\
&\quad \left. -8 \cdot \frac{7.29 + 18.64 \cdot 0.120}{2.34 + 8.49 \cdot 0.120} \sqrt[4]{\frac{0.07}{11.6}} 0.560 \right]
\end{aligned}\tag{7.73}
$$
$$= \frac{1}{\pi^2} 1900 \cdot \frac{18}{0.07} \cdot (223 - 3.54) = 10.86 \cdot 10^6 \, \text{N/m}^2 = 10.86 \, \text{N/mm}^2$$

For the edge $y^b = 0$, the expression Eq. (7.67) for $\sigma_{xx,m}$ is derived and, with Eq. (7.68) and (7.69) respectively, the value of for $\sigma_{xx,m}$ (at $x^b = 0$) becomes

$$\sigma_{xx,m(y^b=0)} = \frac{1}{\pi^2} 1900 \frac{18}{0.07} \cos 0 \left[-\frac{8}{\pi} \frac{18}{11.6} 0.828 \right] \tag{7.74}$$
$$= -0.162 \cdot 10^6 \, \text{N/m}^2 = -0.162 \, \text{N/m}^2$$

The total value of σ_{xx} at the middle of the free edge $(x^b = 0)$ thus is

$$\sigma_{xx(y^b=0)} = \sigma_{xx,b(y^b=0)} + \sigma_{xx,m(y^b=0)} = 10.70 \, \text{N/m}^2 \tag{7.75}$$

The stress found under the assumption that the load can be taken into account as a uniform load, since the shell is shallow, is about 25 % higher than the stress found with the load acting in the vertical direction. The vertical load is more realistic, so replacing by a uniform is safe, but uneconomic. The calculation is easier, but we pay a price.

7.8 Circular Shell Roof Compared with Beam Theory

Figure 7.6 shows the stresses and the normal displacement (in this figure denoted by w) for the shell roof with the geometry of Eq. (7.69) under its own weight of Eq. (7.68). The left column represents the shell supported on two diaphragms with

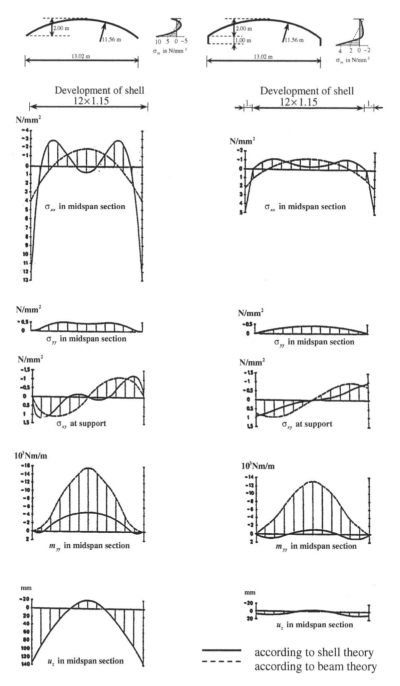

Fig. 7.6 Cylindrical roof with straight edge. Length 18 m, radius 11.56 m, thickness 0.07 m. *Left* without edge beam, *right* with edge beam

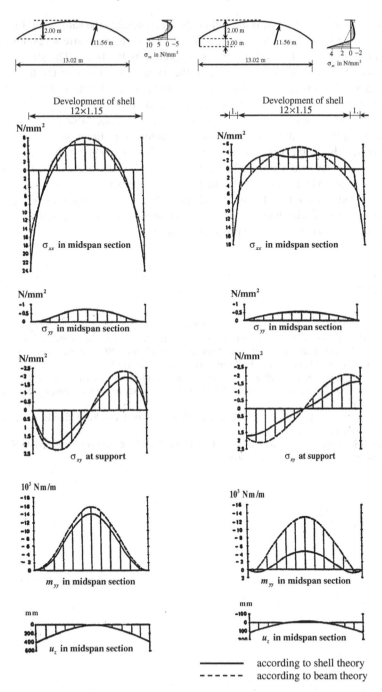

Fig. 7.7 Cylindrical roof with straight edge. Length 36 m, radius 11.56 m. thickness 0.07 m. *Left* without edge beam, *right* with edge beam

straight edges that are free, and the right column represents the same shell but now with edge beams. Because of the edge beams, the membrane stress resultant $n_{x\theta}$ can be present at the edge because $n_{x\theta}$ is supported by the beam. The stresses that are obtained by using an elementary beam theory are also shown in the figure (the dotted lines). The distribution over the cross-section of a beam is linear, but the distribution over the height of the cylindrical roof is not linear. By taking into account the influence of the edge disturbances, we are better able to describe the stress distribution and the deformation in circumferential direction.

Figure 7.7 shows the stresses and the normal displacement for the same cross-sectional geometry and loading, but now for a shell with straight edges with a length of $l = 36$ m instead of $l = 18$ m. The difference between the stresses in this cylindrical roof derived with Tables 7.2 and 7.3 and the stresses obtained with beam theory are smaller because of the larger span between the diaphragms: the longer shell acts more like a simply supported beam.

References

1. Donnell LH (1933) Stability of thin-walled tubes under torsion. NACA Report No 479
2. Bouma AL et al (1956) The analysis of the stress distribution in circular cylindrical shell roofs according the D.K.J. method with aid of an analysis scheme. IBC announcements, Institute TNO for Building Materials and Building Structures, Delft, (in Dutch)
3. Bouma AL, Van Koten H (1958) The analysis of cylindrical shells (in Dutch), Report No. BI-58-4, IBC announcements, Institute TNO for Building Materials and Building Structures, Delft
4. von Karman T, Tsien HS (1941) The Buckling of thin Cylindrical Shells under axial Compression. J Aeronaut Sci 8:303
5. Jenkins RS, (1947) Theory and design of cylindrical shell structures. Report of O.N Arup group of consulting engineers, London

Chapter 8
Hyperbolic- and Elliptic-Paraboloid Roofs

In the previous chapters, we have discussed cylindrical shells, which are special cases of doubly curved shells. We now proceed to more general cases of such shells, the elliptic paraboloid shell and the hyperbolic paraboloid shell. Structural engineers refer to the first category as *elpar* and to the second one as *hyppar*. If built, elpars have a rectangular plan with curved edges (left shell in Fig. 8.1). Hyppars also have rectangular plans, but may have either curved edges (middle shell in Fig. 8.1) or straight edges (right shell in Fig. 8.1). Because hyppars on straight edges are applied most, we will pay most attention to this type, and start with them. At the end of the chapter, the elpar and hyppar with curved edges are addressed only briefly.

The hyppar, a shell with negative Gaussian curvature, is mainly applied for roofs. We will discuss both the membrane state and bending disturbances in edge zones. We restrict the theory to shallow hyppars, so we fall back on the membrane theory of Chap. 3 and the bending theory of Chap. 6.

8.1 Geometry of the Hyppar Surface with Straight Edges

Consider the hyppar of Fig. 8.2. At the centre point of the shell surface, the tangent plane is shown with dotted lines. We choose the set of axes x and y in this plane. The shell surface has zero curvatures in the x- and y-directions, k_x and k_y respectively. The twist k_{xy} is nonzero. In directions 45° with the axes x and y, we observe curvatures; they are equal but have opposite signs. In the one vertical plane over the diagonal the hyppar is concave and in the other convex. In this chapter, we restrict ourselves to the description of the behaviour in the coordinate system of Fig. 8.2. The geometry of the shell surface is defined by

$$z = k_{xy}xy \qquad (8.1)$$

J. Blaauwendraad and J. H. Hoefakker, *Structural Shell Analysis*,
Solid Mechanics and Its Applications 200, DOI: 10.1007/978-94-007-6701-0_8,
© Springer Science+Business Media Dordrecht 2014

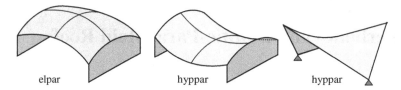

Fig. 8.1 Doubly curved shells of rectangular plan

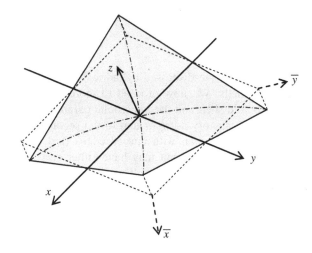

Fig. 8.2 Hyperbolic paraboloid

If we had chosen a description in the set of axes $\bar{x}$ and $\bar{y}$ with an angle of 45°
with the axes x and y, as shown in Fig. 8.1, the geometry equation would be

$$z = \frac{1}{2}k\left(\bar{x}^2 - \bar{y}^2\right) \tag{8.2}$$

The two curvatures $k_{\bar{x}}$ and $k_{\bar{y}}$ in this coordinate system have values k and $-k$,
respectively. We conclude that k is equal to the reciprocal of the radius of the
curved diagonals in Fig. 8.1. If we introduce the symbol a for this radius, we can
write $k = 1/a$. At a closer look, the twist k_{xy} is equal to the (absolute) value of the
curvatures in the lines.

The borders of the surface are of straight lines, but also each intersection of the
surface with planes $x =$ constant and planes $y =$ constant. The shell is an example
of a ruled surface as discussed in Chap. 1. It is a special example, in which two
pairs of straight lines slide over straight generator curves, see Fig. 8.3. The twist
k_{xy} is easily found from the geometry of the shell surface. Accounting for Eq. (8.1),
and defining the lengths l_x and l_y and the elevation f in Fig. 8.4, it holds that

$$f = k_{xy}\, l_x l_y \tag{8.3}$$

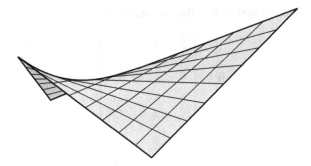

Fig. 8.3 Straight-line generators on a hyppar

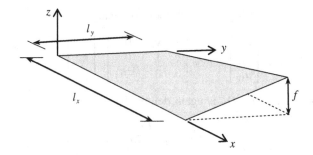

Fig. 8.4 Twisted rectangle with lengths l_x and l_y and elevation f

from which k_{xy} is calculated. The reciprocal value of k_{xy} is called a, which is identified as the radius of curvature over the diagonals of the shell plan. In the chosen co-ordinate system, the hyppar is thus referred to by its generators and can be interpreted as a twisted rectangle.

8.2 Set of Relations for Hyppar with Straight Edges

8.2.1 Kinematic Relation

For the hyppar, it holds that $k_x = 0$, $k_{\bar{y}} = 0$ and $k_{xy} \neq 0$. Then, the kinematic relation of Eq. (3.15) becomes

$$
\begin{bmatrix} \varepsilon_{xx} \\ \varepsilon_{yy} \\ \gamma_{xy} \end{bmatrix} = \begin{bmatrix} \dfrac{\partial}{\partial x} & 0 & 0 \\ 0 & \dfrac{\partial}{\partial y} & 0 \\ \dfrac{\partial}{\partial y} & \dfrac{\partial}{\partial x} & -2k_{yx} \end{bmatrix} \begin{bmatrix} u_x \\ u_y \\ u_z \end{bmatrix}
\tag{8.4}
$$

Symbolically presented as $\mathbf{e} = \mathbf{Bu}$, the matrix $\mathbf{B}$ is

$$\mathbf{B} = \begin{bmatrix} \dfrac{\partial}{\partial x} & 0 & 0 \\[2ex] 0 & \dfrac{\partial}{\partial y} & 0 \\[2ex] \dfrac{\partial}{\partial y} & \dfrac{\partial}{\partial x} & -2k_{xy} \end{bmatrix} \tag{8.5}$$

8.2.2   Constitutive Relation

The constitutive relation in Eq. (3.30) does not change:

$$\begin{bmatrix} n_{xx} \\ n_{yy} \\ n_{xy} \end{bmatrix} = D_m \begin{bmatrix} 1 & v & 0 \\ v & 1 & 0 \\ 0 & 0 & \left(\frac{1-v}{2}\right) \end{bmatrix} \begin{bmatrix} \varepsilon_{xx} \\ \varepsilon_{yy} \\ \gamma_{xy} \end{bmatrix} \tag{8.6}$$

The membrane rigidity D_m is again defined by

$$D_m = \frac{Et}{(1 - v^2)} \tag{8.7}$$

8.2.3   Equilibrium Relation

The equilibrium relation of Eq. (3.27) changes to

$$\begin{bmatrix} -\dfrac{\partial}{\partial x} & 0 & -\dfrac{\partial}{\partial y} \\[2ex] 0 & -\dfrac{\partial}{\partial y} & -\dfrac{\partial}{\partial x} \\[2ex] 0 & 0 & -2k_{xy} \end{bmatrix} \begin{bmatrix} n_{xx} \\ n_{yy} \\ n_{xy} \end{bmatrix} = \begin{bmatrix} p_x \\ p_y \\ p_z \end{bmatrix} \tag{8.8}$$

Symbolically presented as $\mathbf{B}^*\mathbf{n} = \mathbf{p}$, the matrix $\mathbf{B}^*$, the adjoint of $\mathbf{B}$, is thus equal to

$$\mathbf{B}^* = \begin{bmatrix} -\dfrac{\partial}{\partial x} & 0 & -\dfrac{\partial}{\partial y} \\[2ex] 0 & -\dfrac{\partial}{\partial y} & -\dfrac{\partial}{\partial x} \\[2ex] 0 & 0 & -2k_{xy} \end{bmatrix} \tag{8.9}$$

8.3 Membrane Solution for a Uniform Load on Hyppar with Straight Edges

For a uniform normal load p_z, we set $p_x = p_y = 0$; the equilibrium Eq. (8.8) become

$$\frac{\partial n_{xx}}{\partial x} + \frac{\partial n_{xy}}{\partial y} = 0$$

$$\frac{\partial n_{yy}}{\partial y} + \frac{\partial n_{xy}}{\partial x} = 0 \qquad (8.10)$$

$$- 2k_{xy}n_{xy} = p_z$$

These equations imply that a uniform normal load yields a constant membrane force n_{xy}, and therefore the other membrane forces are constant also. We assume boundaries of the hyppar that are not able to support normal forces; at the boundaries only shear membrane forces can be resisted. Because of these conditions, we choose $n_{xx} = n_{yy} = 0$ and Eq. (8.10) yields, accounting from here for $a = k_{xy}^{-1}$,

$$n_{xx} = n_{yy} = 0$$

$$n_{xy} = -\frac{1}{2}ap_z \qquad (8.11)$$

Using the constitutive relation (8.6) in combination with solution (8.11), we can determine the normal strains and the shear strain:

$$\varepsilon_{xx} + \nu\varepsilon_{yy} = 0$$

$$\nu\varepsilon_{xx} + \varepsilon_{yy} = 0 \qquad (8.12)$$

$$D_m \frac{1-\nu}{2}\gamma_{xy} = n_{xy}$$

The normal strains are zero and the shear strain is constant. The strains are thus expressed by

$$\varepsilon_{xx} = \varepsilon_{yy} = 0$$

$$\gamma_{xy} = -\frac{1}{D_m(1-\nu)}ap_z \qquad (8.13)$$

Using the kinematic relations (8.4), we can determine the displacements:

$$\frac{\partial u_x}{\partial x} = 0$$

$$\frac{\partial u_y}{\partial y} = 0 \qquad (8.14)$$

$$\frac{\partial u_x}{\partial y} + \frac{\partial u_y}{\partial x} - \frac{2}{a}u_z = -\frac{1}{D_m(1-\nu)}ap_z$$

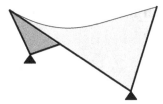

Fig. 8.5 Hyppar with edge beams and two supports

The first two equations show that the tangential displacements u_x and u_y must be constant. For a hyppar supported at the angular points, it follows that these displacements must be zero; the third equation gives

$$u_x = u_y = 0$$
$$u_z = \frac{1}{2D_m(1-\nu)}a^2p_z \tag{8.15}$$

Summing up, the solution for a hyppar subjected to a normal load p_z is

$$n_{xy} = -\frac{1}{2}ap_z$$
$$u_z = \frac{a^2}{2D_m(1-\nu)}p_z \tag{8.16}$$

The normal membrane forces n_{xx} and n_{yy} are zero. Therefore, only a shear force occurs in the hyppar. It means that principal membrane forces will occur under angles of 45°, a tensile membrane force n in the one direction and a compression membrane force $-n$ in the other. The value of n is equal to the absolute value of n_{xy}. At the boundaries, the shear membrane force must be supported, for example, by edge members of the shell. Figure 8.5 shows a hyppar on two supports with four edge members. Each member has a free end and a supported end. The members are supposed to be infinitely rigid in axial direction and have no bending or torsion stiffness. Each edge member is loaded by the shell with a uniform shear membrane force, so a normal force will occur in the member, linearly increasing from zero at the free end of the member to a maximum at the supported member end.

Often the edge members cannot displace in vertical direction because of facades of the building. In that case, the normal displacement u_z of Eq. (8.16) is prevented. As a consequence, an edge disturbance will occur with bending moments and transverse shear forces. This is the subject of Sect. 8.4.

8.3.1 Concluding Remarks About the Membrane Solution

We have derived the membrane solution for the hyperbolic paraboloid shell under a uniform normal load on the basis of three simplifications. The first one regards the shallowness of the shell and states that the elevation is small compared to the span ($f < l$). Then, its own weight can be supposed to act normal to the surface. The principal radii of curvature are in the diagonal direction of the hyppar and have the value a, where twist $a = 1/k_{xy}$.

The second simplification—closely related to the first—involves the loading, which is taken in such a way that the in-plane loads are zero and the normal load is uniformly distributed over the shell surface.

The third simplification involves the boundary conditions, which represent a hyppar that is supported at angular points, and has edge beams with infinite extensional rigidity and zero flexural and torsion rigidities.

For these simplifications the response of the shallow hyppar is rather simple since the membrane solution contains only a constant shear membrane force n_{xy} in the shell and a linearly varying normal force in the edge members. There is only a constant shear strain in the shell and only a constant normal displacement u_z of the middle surface. Principal membrane forces are in the diagonal direction and have the size of the shear force.

8.4 Bending of Hyppar with Straight Edges

8.4.1 Differential Equation

The objective of this chapter is to derive a bending solution for the edge disturbance under the normal load of the hyppar. For this purpose, we can use the differential equation for bending of shallow shells, derived in Chap. 6. As we did for the circular cylindrical shell, we use the membrane solution as the inhomogeneous solution to the differential equation. The homogeneous solution describes the edge disturbance needed to compensate the shortcomings of the membrane response at the boundaries. For this solution, we set the distributed loads on the shell surface zero. In Eq. (6.20) we introduced the shell differential operator Γ and the Laplacian Δ. The curvatures of a hyppar are $k_x = 0$, $k_y = 0$, $k_{xy} = 1/a$, and therefore these operators are

$$\Gamma = -\frac{2}{a}\frac{\partial^2}{\partial x \partial y}$$

$$\Delta = \frac{\partial^2}{\partial x^2} + \frac{\partial^2}{\partial y^2}$$

(8.17)

Because the membrane solution accounts for the inhomogeneous solution to the differential equation, the load terms are set equal to zero. For $p_z = 0$, the homogeneous differential equation of Eq. (6.22) becomes

$$D_b \Delta\Delta\Delta\Delta u_z + D_m (1 - v^2)\Gamma^2 u_z = 0$$

(8.18)

After substitution of the operators in Eq. (8.17), this differential equation is

$$D_b\left(\frac{\partial^2}{\partial x^2}+\frac{\partial^2}{\partial y^2}\right)^4 u_z + D_m\left(1-v^2\right)\frac{4}{a^2}\frac{\partial^4 u_z}{\partial x^2 \partial y^2}=0 \tag{8.19}$$

8.4.2 Approximate Bending Solution for Hyppar with Straight Edges

The exact homogeneous solution to the differential equation (8.18) is rather difficult to derive and therefore it is improbable that it can be made suitable for application in structural design. The objective of this section is to obtain an approximate solution that describes the edge disturbance starting at a straight edge. The approximation is justified by two considerations. Firstly, we assume that an edge load or edge displacement, which is necessary to compensate the shortcomings of the membrane solution, is probably described by a smooth function in the direction of the straight edge. Secondly, we assume that for such an edge load the stress resultants and the displacements vary rapidly in the direction normal to the straight edge. Hence, for an edge normal to the x-direction with length l_y, the derivatives with respect to y are negligibly small compared to those with respect to x and only the highest derivative with respect to x in Eq. (8.18) has to be retained. This implies that all the lower derivatives with respect to x can be neglected in the differential operator $\Delta\Delta\Delta\Delta$ (which is multiplied by the flexural rigidity D_b). Therefore we replace $\Delta\Delta\Delta\Delta$ by $\partial^8/\partial x^8$ .The term with the operator Γ^2 (multiplied by the membrane rigidity D_m) represents the membrane action of the shell, which is considered to be of the same order of magnitude as the bending action. So, this term should not be neglected or adapted. The result is that Eq. (8.19) becomes

$$D_b\frac{\partial^8 u_z}{\partial x^8}+\frac{4D_m\left(1-v^2\right)}{a^2}\frac{\partial^4 u_z}{\partial x^2 \partial y^2}=0 \tag{8.20}$$

Integrating this equation twice with respect to x and dividing by D_b, we obtain

$$\frac{\partial^6 u_z}{\partial x^6}+\frac{4D_m\left(1-v^2\right)}{D_b a^2}\frac{\partial^2 u_z}{\partial y^2}=0 \tag{8.21}$$

It is useful to introduce the parameters λ and α_n:

$$\lambda^4=\frac{4D_m\left(1-v^2\right)}{D_b a^2}=\frac{48\left(1-v^2\right)}{(at)^2};\quad \alpha_n=\frac{n\pi}{l} \tag{8.22}$$

The reduced equation becomes

$$\frac{\partial^6 u_z}{\partial x^6}+\lambda^4\frac{\partial^2 u_z}{\partial y^2}=0 \tag{8.23}$$

For an edge with $x = constant$ the trial solution is

$$u_z(x, y) = Ae^{rx} \cos(\alpha y) \tag{8.24}$$

where $\alpha = \pi/l$, r are the roots to be determined, and A is a constant. In the direction y along the edge, we assume a single cosine wave, because higher wave numbers do not match with the assumption of a smooth function in the direction of the edge. Substitution of this solution in the reduced Eq. (8.21) yields

$$r^6 - \lambda^4 \alpha^2 = 0 \tag{8.25}$$

The additional parameter β is introduced as

$$\beta = \sqrt[6]{\lambda^4 \alpha^2} = \frac{\sqrt[6]{48\pi^2(1 - v^2)}}{\sqrt[3]{(atl)}} \tag{8.26}$$

With this parameter, the reduced equation is

$$r^6 = \beta^6. \tag{8.27}$$

This equation has six roots:

$$r_{1,4} = \pm\beta$$
$$r_{2,5} = \pm\frac{1}{2}\beta\left(1 + i\sqrt{3}\right) \tag{8.28}$$
$$r_{3,6} = \pm\frac{1}{2}\beta\left(-1 + i\sqrt{3}\right)$$

As was the case for a circular cylindrical roof, the function of u_z will have a part that decays for an x-coordinate that starts at the edge, and a part that increases. This is equivalent to a solution part that decays from the considered edge, and a solution part that decays from the opposite edge. We suppose that edges are sufficiently far apart, so both decaying parts have sufficiently vanished at their opposite edge; then only the negative roots are of interest:

$$r_1 = -\beta$$
$$r_2 = -\frac{1}{2}\beta\left(1 + i\sqrt{3}\right) \tag{8.29}$$
$$r_3 = -\frac{1}{2}\beta\left(1 - i\sqrt{3}\right)$$

$$u_z = \left\{ A_1 e^{-\beta x} + e^{-\frac{1}{2}\beta x}\left(A_2 e^{\frac{1}{2}i\sqrt{3}x} + A_3 e^{-\frac{1}{2}i\beta\sqrt{3}x}\right) \right\} \cos(\alpha y) \tag{8.30}$$

The sum and difference of the second and the third root are purely real or purely imaginary and constitute another set of three independent homogeneous solutions. Without loss of generality, we can reshape Eq. (8.30) to

$$u_z = \left\{ A_1 e^{-\beta x} + e^{-\frac{1}{2}\beta x} \left(A_2 \cos\left(\frac{1}{2}\sqrt{3}x\right) + A_3 \sin\left(\frac{1}{2}\sqrt{3}x\right) \right) \right\} \cos(\alpha y) \quad (8.31)$$

The power in the first term is twice the power in the second term; its influence length is thus half that for the second term; therefore we focus on the second term. The terms multiplied by e to the power $-\beta x/2$ have an influence length

$$l_i = \frac{2\pi}{\beta} \approx 2.25 \sqrt[3]{(atl)} \quad (8.32)$$

We may write this as

$$l_i = 2.25 \ l \ \sqrt[3]{\left(\frac{a}{l}\right)^2 \frac{t}{a}} \quad (8.33)$$

For the usual hyppar roof, the thickness-to-radius ratio t/a varies between 1/100 and 1/200 and the radius-to-length ratio a/l varies between 4 and 10. Then the influence length roughly varies between l and $2l$. The outcome l is in accordance with the assumption that the decaying function has vanished at the opposite edge, but the outcome $2l$ is not. Then, at the opposite edge, the decaying function will still be about 20 %, so the solution may be less accurate. However, there will be a reflection of this 20 %, which will become practically zero at the starting edge. We conclude that the derived formulas can be used without concern.

8.4.3 Edge Disturbances for Uniform Load

Loof [1] has developed simplified formulas for the bending disturbance starting at the edges of a hyppar roof under uniform normal load p_z on the basis of the approximate solution (8.31). This has been done for two different boundary conditions, one time for a clamped and one time for a hinged supported edge. We skip the derivations here and limit ourselves to presenting the results. For the clamped edge, we show the distribution of m_{xx} in Fig. 8.6 and for the hinged supported edge in Fig. 8.7. The bending moment on the vertical axis is expressed in terms of the uniform load p, the length of the edge l and the characteristic length $l_c = \sqrt[3]{(atl)}$. The distance on the horizontal axis is non-dimensionalized with $\sqrt[3]{(atl)}$. Of special interest is the value of the bending moment m_{xx} and the transverse shear force v_x at the edge. This edge, for example, can be connected to a solid wall or an edge beam. For convenience, we introduce

$$l_c = \sqrt[3]{(atl)}, \quad \xi = l_c/l \quad (8.34)$$

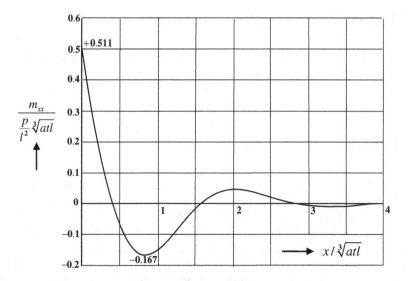

Fig. 8.6 Edge disturbance for m_{xx} in a hyppar with clamped edges

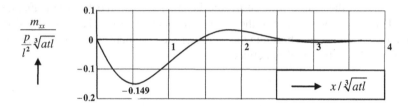

Fig. 8.7 Edge disturbance for m_{xx} in a hyppar with hinged supported edges

For the clamped edge (Fig. 8.6), we have the following results:

$$\begin{aligned} x = 0 \qquad & m_{xx} = 0.511 \ \xi^4 pl^2 \\ & v_x = 1.732 \ \xi pl \\ x = 0.85 \ l_c \quad & m_{xx} = -0.167 \ \xi^4 pl^2 \end{aligned} \tag{8.35}$$

For the hinged supported edge (Fig. 8.7), we have

$$\begin{aligned} x = 0.55 \quad & l_c m_{xx} = -0.149 \ \xi^4 pl^2 \\ x = 0; \quad & v_x = 0.577 \ \xi pl \end{aligned} \tag{8.36}$$

Both for the clamped edge and the hinged edge, the shear force is maximal at the edge ($x = 0$). At a clamped edge, the maximum bending moment occurs at the edge ($x = 0$) with tensile stresses at the top face of the roof shell. A smaller extreme bending moment of opposite sign occurs at a distance $x = 0.85 \ l_c$ of the edge. At a hinged edge the maximum bending moment occurs inward on the shell at a distance $x = 0.55 \ l_c$ from the edge, and raises tensile stresses at the bottom face of the roof shell.

8.5 Hyppar Roof Examples

We will apply the theory of the previous sections to two roof structures. First we consider a single hyppar on two supports as shown in Fig. 8.5. Next we calculate in Sect. 8.5.2 the stress state in a roof that is composed of four hyppars.

8.5.1 Single Hyppar on Two Supports

Consider the structure of reinforced concrete of Fig. 8.8 with a square plan. We bring in edge members along all four edges. The shell has two supporting blocks at the ends of a diagonal of the plan. The centre of the block coincides with the intersection of two edge members. A tension tie connects the two blocks. The figure shows an exploded view of the shell in which the edge members are separated from the shell. It is shown which shear force n_{xy} (in the figure denoted by n) is acting on the members and which one on the shell. The normal force in the member increases linearly from zero to the maximum value N. All member forces are compression forces.

We choose the length $a = 24$ m, the height $f = 4$ m and the shell thickness $t = 0.1$ m. The cross-section of the edge members is 0.15 m^2. The axes x and y to define the geometry are chosen in the centre of the base plan. The z-axis is upward. The geometry of the shell is defined by

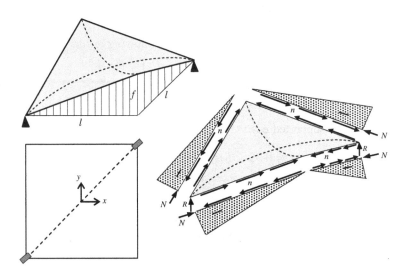

Fig. 8.8 Roof of a single hyppar on two supports

$$z(x,y) = \left(\frac{1}{2} + \frac{4xy}{l^2}\right)f \tag{8.37}$$

We calculate the shell for a combination of its own weight and a layer of snow, for which we take $p = 5000$ N/m^2. This number includes roof covering, finish, snow and partial load factors. The shell is shallow, so we consider the load p to act normal to the shell surface. Its own weight of the edge members is not considered. We choose the following material properties. Young's modulus of concrete is $E = 2 \times 10^4$ N/mm^2. Poisson's ratio is taken to be zero.

Calculation of Stresses and Moments

From Eq. (8.37), we calculate

$$k_{xy} = \frac{1}{a} = \frac{\partial^2 z}{\partial x \partial y} = \frac{2f}{l^2} = \frac{2 \times 4}{24^2} = \frac{1}{72} \text{ m}^{-1}.$$

The membrane force is

$$n_{xy} = \frac{1}{2}pa = \frac{1}{2} \times 5000 \times 72 = 180,000 \text{ N/m} = 180 \text{ N/mm}.$$

The shear stress is

$$\sigma_{xy} = \frac{n_{xy}}{t} = \frac{180}{100} = 1.80 \text{ N/mm}^2.$$

Bending Moments

The characteristic parameter is

$$l_c = \sqrt[3]{atl} = \sqrt[3]{72 \times 0.1 \times 24} = 5.57 \text{ m}$$

and the dimensionless parameter ξ is

$$\xi = \frac{l_c}{l} = \frac{5.57}{24} = 0.232, \qquad \xi^4 = 0.00292$$

Close to the blocks, the edge may behave as a clamped edge. Then the bending moment at the edge ($x = 0$) is

$$m_{xx} = 0.511 \, \xi^4 pl^2 = 0.511 \times 0.00292 \times 5000 \times 24^2 = 4297 \text{ Nm/m}$$

which needs top reinforcement in the shell.

Halfway along the edge member, the state is to be compared with a hinged edge. There the bending moment is

$$m_{xx} = -0.149 \, \xi^4 pl^2 = -0.149 \times 0.00292 \times 5000 \times 24^2 = -1253 \text{ Nm/m}$$

at a distance $x = 0.55\, l_c = 0.55 \times 5.57 = 3.1$ m. This moment requires bottom reinforcement.

Stress in Edge Members

The maximum value of the normal force N in the edge members is $N = n_{xy} l^*$, $l^* = l/\cos\alpha$.

Herein l^* is the length of the edge member and α is the angle between the edge member and the projection of the member on the x, y-plane. The angle is calculated from $\tan\alpha = f/l$. Then $l^* = \sqrt{l^2 + f^2}$. For our structure, we obtain $l^* = \sqrt{24^2 + 4^2} = 24.3$ m. Then N becomes $N = 180{,}000 \times 24.3 = 4.37 \times 10^6$ N. With $A = 0.15$ m^2, we obtain

$$\sigma = \frac{4.37 \times 10^6}{0.15 \times 10^6} = 29.2 \text{ N/mm}^2.$$

Force in Tension Tie

The decomposition of the member normal force leads to a vertical component $N_v = N\sin\alpha$ and horizontal component, which is equal to $n_{xy} l$. Therefore, $N_h = n_{xy} l = 180{,}000 \times 24 = 4.32 \times 10^6$ N. The tensile force T in the cable between the blocks is $\sqrt{2}$ times larger,

$$T = \sqrt{2} \times 4.32 \times 10^6 = 6.11 \times 10^6 \text{ N}.$$

Overall Checks

The first check is that the sum of the vertical components of the four edge member forces N must be equal to the total load p. The sum V of the vertical components is $V = 4N_v = 4N\sin\alpha$. We know $N = n_{xy} l^*$, $\sin\alpha = f/l^*$, $n_{xy} = pa/2$ and $k_{xy} = 1/a = 2f/l^2$. Therefore, $V = pl^2$, which indeed is the load on the shell.

The second check regards moments, for which we consider Fig. 8.9. We calculate the external moment M_{ext} due to the distributed load on the triangular half shell, and the internal global moment M_{int} due to the principal membrane force n and the member forces N.

The load on the top half triangular shell part left of the diagonal raises a moment M_{ext} about the diagonal (left part of Fig. 8.9) which must be in equilibrium with the moment M_{int} of the membrane tension forces n in the shell section over the diagonal (right part of Fig. 8.9).

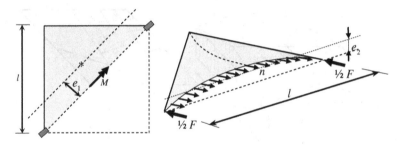

Fig. 8.9 Global equilibrium check

The moment $M_{ext} = Pe_1$, where $P = pl^2/2$ is the load on the triangle shell part, and the eccentricity $e_1 = \frac{1}{3}(l/\sqrt{2})$ defines the position of the centre of gravity of the triangle. So,

$$M_{ext} = \left(\frac{1}{2}pl^2\right)\left(\frac{1}{3}\frac{l}{\sqrt{2}}\right) = \frac{\sqrt{2}}{12}pl^3.$$

Homogenous tensile membrane forces n occur in the parabolic shell section, where $n = n_{xy}$. Because of the shallowness of the shell, the developed length of the parabola can be put equal to the length $l\sqrt{2}$ of the diagonal. Therefore, the resultant of the tensile forces is $N = n\,l\sqrt{2}$. This resultant is at a distance e_2 from the bottom plane, which is two-thirds of the depth $f/2$ of the parabola. So, $e_2 = f/3$. The resultant N of the forces n is balanced by compression forces $\frac{1}{2}N$ in the two blocks. The tensile force N and the two compression forces $\frac{1}{2}N$ are at a distance e_2, so they yield a moment $M_{int} = Ne_2$. Accounting for $N = n_{xy}l\sqrt{2}$, $n_{xy} = \frac{1}{2}pa$, $a = k_{xy}^{-1} = \frac{l^2}{2f}$ and $e_2 = f/3$, we obtain

$$M_{int} = \left(\frac{1}{2}p\right)\left(\frac{l^2}{2f}\right)\left(l\sqrt{2}\right)\left(\frac{1}{3}f\right) = \frac{\sqrt{2}}{12}pl^3.$$

We conclude that the moments M_{ext} and M_{int} are equal, so global equilibrium is satisfied.

8.5.2 Composed Hyppar Roofs

Consider the two structures of reinforced concrete of Fig. 8.10 with a rectangular plan, a building roof and a pavilion roof. The lengths in the plan are $2l_1$ and $2l_2$. Each roof is composed of four hyppars. Edge members occur along all outer edges and in the interface of hyppars. The building roof is supported at the four corners, the pavilion roof shell. Tension ties are needed in the building roof, see the dotted lines. They do not occur in the pavilion roof. The figure shows an expanded view

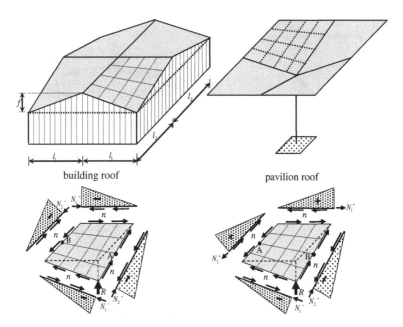

Fig. 8.10 Roofs composed of four hyppars

of one hyppar of each structure, the one with drawn generators. In the figure, the membrane shear force is indicated by n as done before in Fig. 8.8. The normal force in the member increases linearly from zero to the maximum value. In the building roof all members are in compression; then we call the maximum member forces N_1^- and N_2^-. In the pavilion roof, both tension and compression forces occur, with maxima N_1^+, N_2^+, N_1^- and N_2^-.

We choose lengths $l_1 = 16$ m and $l_2 = 24$ m, height $f = 4$ m and the shell thickness $t = 0.1$ m. The cross-section of the outer edge members is 0.125 m^2 and of the inner members is 0.25 m^2. We calculate the shell for a combination of its own weight and a layer of snow, for which we take $p = 4400$ N/m^2. This number includes roof covering, finish, snow and partial load factors. The shell is shallow, so we consider the load p to act normal to the shell surface. Their own weight of the edge members is not considered. We choose the following material properties. Young's modulus of concrete is $E = 2 \times 10^4$ N/mm^2. Poisson's ratio is taken to be zero. After the extensive treatment of the example in the previous section, we only briefly reproduce the figures.

Calculation of Stresses and Moments

$$k_{xy} = \frac{f}{l_1 l_2} = \frac{4}{16 \times 24} = \frac{1}{96} \text{ m}^{-1}; \quad a = \frac{1}{k_{xy}} = 96 \text{ m}.$$

The membrane force is

$$n_{xy} = \frac{1}{2}pa = \frac{1}{2} \times 4,400 \times 96 = 211,200 \text{ N/m} = 211.2 \text{ N/mm}.$$

The shear stress is

$$\sigma_{xy} = \frac{n_{xy}}{t} = \frac{211}{100} = 2.11 \text{ N/mm}^2.$$

Bending Moment

The characteristic parameter is $l_c = \sqrt[3]{atl}$, where l is the length of the considered edge. The larger the length l, the larger the edge moment is. Therefore we use edge length l_2:

$$l_c = \sqrt[3]{atl_2} = \sqrt[3]{72 \times 0.1 \times 24} = 5.57 \text{ m}$$

and the dimensionless parameter ξ is

$$\xi = \frac{l_c}{l_2} = \frac{5.57}{24} = 0.232, \qquad \xi^4 = 0.00290.$$

At point A of the shells, the bending moment is (bottom reinforcement)

$$m_{xx} = -0.149\,\xi^4 pl^2 = -0.149 \times 0.00290 \times 4400 \times 24^2 = -1095 \text{ Nm/m}.$$

At point B of the shells, the bending moment is (top reinforcement)

$$m_{xx} = 0.511\,\xi^4 pl_2^2 = 0.511 \times 0.0029 \times 4400 \times 24^2 = 3756 \text{ Nm/m}.$$

The reinforcement for this moment is needed over a length of $3\,l_c = 16.7$ m, which is two-thirds of the shell length. Furthermore, this bottom reinforcement is needed from two opposite edges, so, in fact over the full shell area.

Stress in Edge Members

The largest force will occur in members with length l_2^*. The maximum value of the normal force N_2 in the edge members is

$$N_2 = n_{xy}l_2^*, \qquad l_2^* = l_2/\cos\alpha.$$

Herein l_2^* is the length of the edge member and α is the angle between the edge member and the projection of the member on the x, y-plane. The angle is calculated from $\tan\alpha = f/l_2$. Then $l_2^* = \sqrt{l_2^2 + f^2}$. For our structure, we obtain $l_2^* = \sqrt{24^2 + 4^2} = 24.331$ m. Then N_2 becomes $N_2 = 211.200 \times 24.331 = 5.139 \times 10^6$ N. With $A = 0.125$ m^2, we obtain

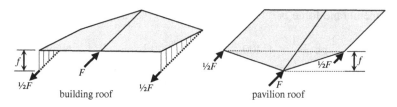

Fig. 8.11 Global equilibrium check

$$\sigma = \frac{5.139 \times 10^6}{0.125 \times 10^6} = 41.1 \text{ N/mm}^2.$$

Force in Tension Tie

For the tie with length l_1, it holds that

$$N_{1h} = n_{xy}l_1 = 211.200 \times 16 = 3.379 \times 10^6 \text{ N}.$$

For the tie with length l_2, it holds that

$$N_{2h} = n_{xy}l_2 = 211.200 \times 24 = 5.069 \times 10^6 \text{ N}.$$

Overall Check

We will check the overall equilibrium of roof halves, for which we consider
Fig. 8.11. We call the global moment due to the internal forces M_{int} and to the
external load M_{ext}. The total load P on the roof half is $P = p(2l_1)\, l_2 = 2p\, l_1\, l_2$.
The lever arm e of this load is $e = l_2/2$. Therefore the global moment due to the
distributed load p is

$$M_{ext} = Pe = (2pl_1l_2)(l_2/2) = pl_1\, l_2^2.$$

The internal global moment is due to the forces F and $2F$ in the edge members.
The force F is the horizontal component N_{2h} in l_2-direction:

$$F = N_{2h} = n_{xy}l_2.$$

The global moment is

$$M_{int} = (2F)f = (2n_{xy}l_2)f = \left(2\left(\frac{1}{2}pa\right)l_2\right)f = pal_2f.$$

Accounting for $a = l_1\, l_2/f$, the global moment is $M_{int} = p\, l_1\, l_2^2$. We conclude
that the moments M_{int} and M_{ext} are equal, so global equilibrium is satisfied.

8.6 Elpars and Hyppars with Curved Edges

Shortly, we will address the item of elpars and hyppars of a rectangular plan with curved edges. We will not derive here a full membrane and bending solution; instead it will suffice to state the governing differential equation and present some characteristic results from Bouma [2]. Consider the roof shells of Fig. 8.12. The two left shells are elpars and the two right ones hyppars. The transitions in between are cylinders. Of these six shells, the three top ones are supported in two opposite edges. The supporting walls are supposed to be infinitely rigid in-plane and very flexible for displacements and rotations out-of-plane. The shell is fixed shear-stiff to the wall. The two longest edges of these shells are free. Therefore, edge moments in the three shells are zero at all edges. The shells behave as 'beams' in x-direction. We will address the left shells as 'elpar beam', the middle shells as 'cylinder beam', and the right shells as 'hyppar beam'.

The three shells in the bottom part of the figure are shear-stiff supported along all four edges. This raises expectations of much more reduced displacements, membrane forces and moments. Again, edge moments in these shells are zero at all edges.

A set of axes is chosen such that the x-axis is in the direction of the longest edge, the y-direction in the shortest direction and the z-axis downward. All six shells have the same rectangular plan ($a = 18$ m and $b = 13$ m) and the same radius of curvature in y-direction ($r_y = 11.56$ m). The radius of curvature in x-direction varies, because different values of the elevation (or sag) h are considered, 200 cm, 0 cm, -50 cm and -200 cm, respectively. In all six shells $k_x \neq 0$, $k_y \neq 0$ and $k_{xy} = 0$. Differential equation (6.22) applies again,

$$D_b \Delta\Delta\Delta\Delta u_z + D_m (1 - v^2) \Gamma^2 u_z = \Delta\Delta p_z \qquad (8.38)$$

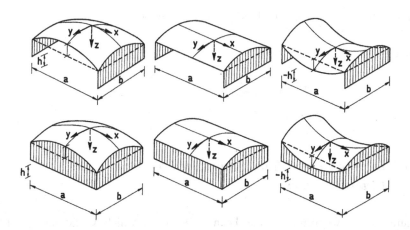

Fig. 8.12 Six doubly curved shells of different shape and way of support

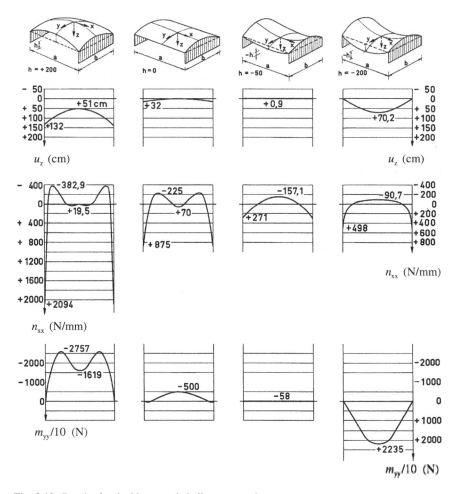

Fig. 8.13 Results for doubly curved shells on two edge supports

where now

$$\Gamma = k_x \frac{\partial^2}{\partial y^2} + k_y \frac{\partial^2}{\partial x^2}; \quad \Delta = \frac{\partial^2}{\partial x^2} + \frac{\partial^2}{\partial y^2} \tag{8.39}$$

8.6.1 Doubly Curved Shells Supported on Two Opposite Edges

Figure 8.13 shows results for the 'beam type' shells, which we extract from [2]. We only changed notation and transformed to modern units. The plot shows deflections u_z, the membrane force n_{xx} and the moment m_{yy} in the middle section of the shell at $x = 0$. Notice that the 'hyppar beam' with $h = -50$ cm is very stiff (very small u_z), has a negligible bending moment m_{yy}, and a distribution of n_{xx} over the shell section which is equal to beam theory. This doubly curved shell behaves exactly as a beam. The maximum membrane force n_{xx} in the 'cylinder beam' ($h = 50$ cm) is more than three times larger. In the 'elpar beam' with $h = 200$ cm the difference is about eight times. Now the behaviour is far from beam theory. The 'hyppar beam' with $h = -200$ cm has a maximum membrane force which is less than two times the force of ideal beam theory.

We conclude that the 'elpar beam' and 'hyppar beam' (each with 200 cm) result in very different values of the membrane force n_{xx}. This is not true for the bending moment m_{yy}. Then values of the same order of magnitude occur, be it of different sign.

On the basis of this comparison it appears that 'beam type' doubly curved shells have very reduced stiffness for increasing elevation and sag. For the considered shells it is recommended to restrict to $-50 < h < 50$ cm.

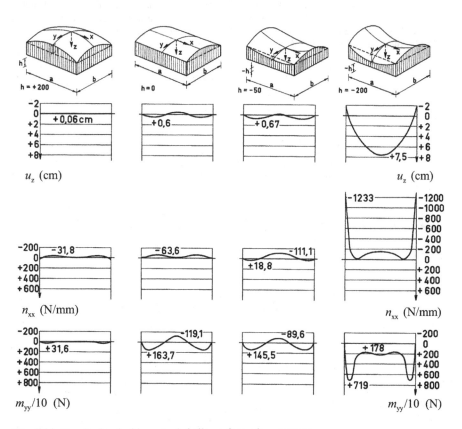

Fig. 8.14 Results for doubly curved shells on four edge supports

8.6.2 Doubly Curved Shells Supported Along All Edges

Figure 8.14 depicts the results of the doubly curved shells, which are supported along the four edges. The plots speak for themselves if they are compared with the plots of the previous figure. Now the shell becomes stiffer the larger the elevation h is. For $h > -50$ cm always very stiff behaviour is found, both for small and large h. Strongly reduced values of the membrane force and bending moment occur, compared to the shells on two edge supports. While the elpar with $h = 200$ cm on two edge supports was the worst case in Fig. 8.13, it is the best one in Fig. 8.14. For four edge supports, the state of stress becomes rapidly unpleasantly large for $h < -50$ cm, and the deflection increases dramatically.

References

1. Loof HW (1961) Edge disturbances in a hyppar shell with straight edges, Report 8-61-3-hr-1. Stevin Laboaratory, Department of Civil Engineering, Technical University Delft (in Dutch)
2. Bouma AL (1960) Some applications of the bending theory regarding doubly curved shells. In: Proceedings of the Symposium on theory of thin shells, North Holland Publishing Company, Delft, August 1959, pp 202–235

Part III
Chimneys and Storage Tanks

Chapter 9
Morley Bending Theory for Circular Cylindrical Shells

The Donnell theory of shallow shells does not accurately apply to fully-closed circular cylindrical shells like chimneys and storage tanks. The main reason is the simplifying assumption that we can use the formulas of the flat plate theory for the change of curvatures in the shell. In full circular cylindrical shells, this is not sufficiently accurate, because then rigid body motions would lead to non-zero changes of curvatures. Especially for structures like long industrial chimneys, storage tanks and pipelines, the imperfections are undesirable and result in substantial errors.

It is generally agreed that a complete and adequate set of equations for the theory of thin shells was developed by Love [1]. Koiter, one time, paraphrased a popular Beatles song: 'All you need is Love, Love is all you need'. More or less equally complicated sets of equations were published later on by, among others, Flügge [2], Wlassow [3], Novoshilov [4], Niordson [8] and Reissner [5]. It appears that just one rigorous theory does not exist. For the special case of circular cylindrical shells, Morley has published an elegant theory [6], which is equally as accurate as Flügge's much more complicated equations. He has suggested his equation on reasoning and judgement, and Koiter (and others after him) have derived it more explicitly. In the present chapter, we will describe this *Morley theory*. The application to chimneys in Chap. 12 and storage tanks in Chap. 13 is taken from Hoefakker [7].

9.1 Introduction

We restrict the applications in the present chapter to cases in which only loading occurs normal to the shell surface for convenience and without degenerating the generality of the approach. Therefore we put $p_x = 0$ and $p_\theta = 0$. Furthermore, the nonzero load p_z is symmetric with respect to the axis $\theta = 0$, and we will restrict ourselves to a load which is constant or linear in axial direction. We consider full circular cylindrical shells, for which the boundary conditions are specified at circular edges.

J. Blaauwendraad and J. H. Hoefakker, *Structural Shell Analysis*,
Solid Mechanics and Its Applications 200, DOI: 10.1007/978-94-007-6701-0_9,
© Springer Science+Business Media Dordrecht 2014

9.1.1 Leading Term in Differential Equations

The changes to the Donnell bending theory of Chap. 7, as required to arrive at the Morley bending theory, are introduced without elaboration. In the next chapters, we will provide and discuss a comparison between results obtained with the Morley bending theory and those obtained with the Donnell bending theory. The eighth-order Donnell equation for circular cylindrical shells reads:

$$\Delta\Delta\Delta\Delta u_z + 4\beta^4 \frac{\partial^4 u_z}{\partial x^4} = \frac{1}{D_b}\Delta\Delta p_z \tag{9.1}$$

in which Δ is the Laplacian:

$$\Delta = \frac{\partial^2}{\partial x^2} + \frac{1}{a^2}\frac{\partial^2}{\partial\theta^2}. \tag{9.2}$$

The definition of the parameter β is

$$\beta^4 = \frac{3(1-v^2)}{(at)^2}. \tag{9.3}$$

It can be concluded that, since $t < a$ for any circular cylinder, the dimensionless parameter $\beta a > 1$. Therefore, in case that derivatives of the fourth order with respect to x of the function u_z exist, the one that is multiplied with the parameter $(\beta a)^4$ will be a leading term since $(\beta a)^4 \gg 1$. In the Morley theory profit is taken from this information.

9.1.2 Geometrical Considerations

The geometry and coordinate system for the circular cylindrical shell are introduced in Sect. 7.1. It holds that

$$k_x = 0; \quad k_y = -1/a; \quad k_{xy} = 0. \tag{9.4}$$

Again, we adopt the vectors in Eq. (7.1) to describe the kinematical, constitutive and equilibrium relations (for loads p_z only):

$$\begin{aligned}
\mathbf{u} &= \begin{bmatrix} u_x & u_\theta & u_z \end{bmatrix}^T \\
\mathbf{e} &= \begin{bmatrix} \varepsilon_{xx} & \varepsilon_{\theta\theta} & \gamma_{x\theta} & \kappa_{xx} & \kappa_{\theta\theta} & \rho_{x\theta} \end{bmatrix}^T \\
\mathbf{s} &= \begin{bmatrix} n_{xx} & n_{\theta\theta} & n_{x\theta} & m_{xx} & m_{\theta\theta} & m_{x\theta} \end{bmatrix}^T \\
\mathbf{p} &= \begin{bmatrix} 0 & 0 & p_z \end{bmatrix}^T.
\end{aligned} \tag{9.5}$$

The scheme of relationships of Fig. 9.1 applies.

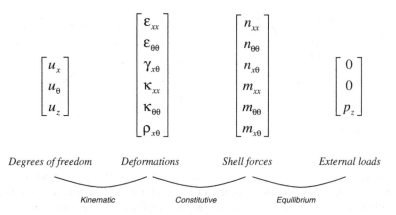

Fig. 9.1 Scheme of relationships for a circular cylindrical shell

9.1.3 Load Considerations

Because of symmetry of the load p_z in circumferential direction with respect to the axis $\theta = 0$, the load is an even periodic function with period 2π with respect to the line of symmetry. The Fourier series of any even function consists only of the even trigonometric functions $\cos n\theta$, and a constant term:

$$p_z(x, \theta) = \sum_{n=0}^{\infty} p_{zn}(x) \cos n\theta, \tag{9.6}$$

where n is the circumferential mode number representing the number of whole waves in circumferential direction. The reader can easily extend the application to an asymmetric load by describing combinations of sine and cosine series per load term.

9.1.4 Three Load-Deformation Behaviours

The behaviour of circular cylindrical shells under the above-defined load is excellently described by Morley's equation, where all quantities can be expressed as functions of the type

$$\phi(x, \theta) = \sum_{n=0}^{\infty} \phi_n(x) \cos n\theta \tag{9.7}$$

and

$$\phi(x, \theta) = \sum_{n=0}^{\infty} \phi_n(x) \sin n\theta \tag{9.8}$$

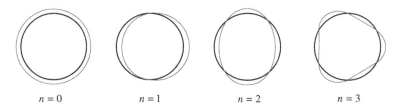

$n = 0$ $n = 1$ $n = 2$ $n = 3$

Fig. 9.2 Four load terms and deformation modes for a *ring*

depending on the axis of symmetry of the quantity under consideration. Nevertheless, we can subdivide the response of a cylinder to all possible loads indicated by a different mode number n into three different load-deformation behaviours. Consider a (long) circular cylinder without restricting boundary conditions and subject to a constant load in x-direction. The response of such a cylinder is equal to the response of a ring to that load. In Fig. 9.2 the load and the corresponding deformation for four terms is displayed for a circular ring.

Axisymmetric Mode

The mode indicated by $n = 0$ (left in Fig. 9.2) is generally known as the *axisymmetric mode* and describes a constant behaviour in circumferential direction. Such a load leads, in principle, to a change in the radius of the cylinder with circular edges. Any quantity ϕ must be constant in circumferential direction; in other words, the substitution $\partial \phi / \partial \theta = 0$ is to be made in the governing equations. All displacements, rotations, membrane forces, moments and shear forces are zero, except u_z, $n_{\theta\theta}$, m_{xx} and v_x. In case of nonzero lateral contraction, also a nonzero bending moment $m_{\theta\theta}$ occurs equal to v times m_{xx}. In Sect. 9.9, we will make an assessment of the relation of the results for the Morley equation to the results as obtained in Chap. 5, which in fact can be considered as results of the Donnell equation.

Beam Mode

The mode indicated by $n = 1$ (second left in Fig. 9.2) is generally known as the *beam mode* and describes the response of the circular cylinder that is obtained if we treat it as a beam with a circular cross-section. In other words, the deflection of the circular cylinder is caused by the resultant of the load term. We must be aware that using the expressions derived for three independent shell displacements (u_x, u_θ and u_z) accounts for both flexural and shear rigidity. Moreover, we obtain a solution that takes care of nonconforming deformation states at the circular boundary. This is due to the fact that the governing equation is an eighth-order differential equation and not only the fourth-order polynomial solution representing the response of Euler beam theory. We conclude that this part of the solution describes an edge disturbance that mainly originates from constrained cross-sectional deformation. Morley's equation excellently fits this behaviour.

For $n = 1$, all quantities can be expressed as functions of the type

$$(x, \theta) =_1 (x) \cos \theta \tag{9.9}$$

and

$$\phi(x, \theta) = \phi_1(x) \sin \theta \tag{9.10}$$

depending on the axis of symmetry of the quantity under consideration. In Sect. 9.9 we will make the relation of the Morley results to the results as obtained in Chap. 4 for the membrane theory only.

Self-Balancing Modes

The modes indicated by $n = 2, 3, 4, \ldots$ ($n = 2$ and $n = 3$ are depicted at the right-hand side in Fig. 9.2) are generally known as the *self-balancing modes*. The load has as many symmetry axes as the mode number n. These axes cross each other at the middle point of the circle, which also holds for n anti-symmetry axes. The response of a ring to such a load is fully described by a deformation of the circular shape without displacing the middle point of that circle since the resultant of the load is equal to zero.

The response of a full cylinder without restriction to the deformation at its circular boundaries will be equal to the response of a ring with the circular profile. Only the membrane force, bending moment and transverse shear force in circumferential direction will occur. However, if this response behaviour is restrained at any circular edge, also bending and membrane straining in axial direction is provoked. Morley's equation excellently describes this behaviour. For mode numbers $n > 1$, all quantities can be expressed as functions of the type

$$\phi(x, \theta) = \sum_{n=2}^{\infty} \phi_n(x) \cos n\theta \tag{9.11}$$

and

$$\phi(x, \theta) = \sum_{n=2}^{\infty} \phi_n(x) \sin n\theta \tag{9.12}$$

depending on the axis of symmetry of the quantity under consideration.

9.2 Sets of Equations

9.2.1 Kinematical Relation

With reference to [8], we introduce the improved expressions for the changes of curvatures $\kappa_{\theta\theta}$ and $\rho_{x\theta}$:

$$\kappa_{\theta\theta} = -\frac{1}{a^2}\frac{\partial^2 u_z}{\partial\theta^2} - \frac{u_z}{a^2}$$

$$\rho_{x\theta} = \frac{2}{a}\frac{\partial u_\theta}{\partial x} - \frac{2}{a}\frac{\partial^2 u_z}{\partial x\partial\theta}$$

(9.13)

which are zero for rigid body motions. Therefore, we upgrade the kinematical relation $\mathbf{e} = \mathbf{Bu}$ of Eq. (6.5) to

$$
\begin{bmatrix}
\varepsilon_{xx} \\
\varepsilon_{\theta\theta} \\
\gamma_{x\theta} \\
\kappa_{xx} \\
\kappa_{\theta\theta} \\
\rho_{x\theta}
\end{bmatrix}
=
\begin{bmatrix}
\dfrac{\partial}{\partial x} & 0 & 0 \\[2mm]
0 & \dfrac{1}{a}\dfrac{\partial}{\partial\theta} & \dfrac{1}{a} \\[2mm]
\dfrac{1}{a}\dfrac{\partial}{\partial\theta} & \dfrac{\partial}{\partial x} & 0 \\[2mm]
0 & 0 & -\dfrac{\partial^2}{\partial x^2} \\[2mm]
0 & 0 & -\dfrac{1}{a^2}\dfrac{\partial^2}{\partial\theta^2} - \dfrac{1}{a^2} \\[2mm]
0 & \dfrac{2}{a}\dfrac{\partial}{\partial x} & -\dfrac{2}{a}\dfrac{\partial^2}{\partial x\partial\theta}
\end{bmatrix}
\begin{bmatrix}
u_x \\
u_\theta \\
u_z
\end{bmatrix}.
$$

(9.14)

9.2.2 Constitutive Relation

The constitutive relation of Eq. (6.6) is still valid:

$$
\begin{bmatrix}
n_{xx} \\
n_{\theta\theta} \\
n_{x\theta} \\
m_{xx} \\
m_{\theta\theta} \\
m_{x\theta}
\end{bmatrix}
=
\begin{bmatrix}
D_m & \upsilon D_m & 0 & 0 & 0 & 0 \\
\upsilon D_m & D_m & 0 & 0 & 0 & 0 \\
0 & 0 & D_m\frac{1-\upsilon}{2} & 0 & 0 & 0 \\
0 & 0 & 0 & D_b & \upsilon D_b & 0 \\
0 & 0 & 0 & \upsilon D_b & D_b & 0 \\
0 & 0 & 0 & 0 & 0 & D_b\frac{1-\upsilon}{2}
\end{bmatrix}
\begin{bmatrix}
\varepsilon_{xx} \\
\varepsilon_{\theta\theta} \\
\gamma_{x\theta} \\
\kappa_{xx} \\
\kappa_{\theta\theta} \\
\rho_{x\theta}
\end{bmatrix}.
$$

(9.15)

The membrane rigidity D_m and flexural rigidity D_b are the same as in the Donnell theory:

$$D_m = \frac{Et}{1-\upsilon^2}; \qquad D_b = \frac{Et^3}{12(1-\upsilon^2)}.$$

(9.16)

The constitutive relation is based on a linear description of the stress distribution across the thickness. We obtain the respective normal stresses and shear stress conveniently from relations, which are equal to those for a thin plate:

$$\sigma_{xx} = \frac{n_{xx}}{t} + z\frac{12m_{xx}}{t^3}$$

$$\sigma_{\theta\theta} = \frac{n_{\theta\theta}}{t} + z\frac{12m_{\theta\theta}}{t^3} \qquad (9.17)$$

$$\sigma_{x\theta} = \frac{n_{x\theta}}{t} + z\frac{12m_{x\theta}}{t^3}.$$

9.2.3 Equilibrium Relation

To arrive at the equilibrium relation $\mathbf{B}^*\mathbf{s} = \mathbf{p}$, we make use of the fact that the differential operator matrix $\mathbf{B}^*$ is the adjoint of the differential operator matrix $\mathbf{B}$ in the kinematical relation Eq. (9.14). The equilibrium relation (6.10) is then upgraded to

$$\begin{bmatrix} -\dfrac{\partial}{\partial x} & 0 & -\dfrac{1}{a}\dfrac{\partial}{\partial\theta} & 0 & 0 & 0 \\[2mm] 0 & -\dfrac{1}{a}\dfrac{\partial}{\partial\theta} & -\dfrac{\partial}{\partial x} & 0 & 0 & -\dfrac{2}{a}\dfrac{\partial}{\partial x} \\[2mm] 0 & \dfrac{1}{a} & 0 & -\dfrac{\partial^2}{\partial x^2} & -\dfrac{1}{a^2}\dfrac{\partial^2}{\partial\theta^2} - \dfrac{1}{a^2} & -\dfrac{2}{a}\dfrac{\partial^2}{\partial x\partial\theta} \end{bmatrix}$$

$$\cdot \begin{bmatrix} n_{xx}a \\ n_{\theta\theta}a \\ n_{x\theta}a \\ m_{xx}a \\ m_{\theta\theta}a \\ m_{x\theta}a \end{bmatrix} = \begin{bmatrix} 0 \\ 0 \\ p_z a \end{bmatrix}. \qquad (9.18)$$

The transverse shear forces, described by Eq. (6.9), do not change:

$$v_x = \frac{\partial m_{xx}}{\partial x} + \frac{1}{a}\frac{\partial m_{x\theta}}{\partial\theta}; \qquad v_\theta = \frac{1}{a}\frac{\partial m_{\theta\theta}}{\partial\theta} + \frac{\partial m_{x\theta}}{\partial x}. \qquad (9.19)$$

9.2.4 Boundary Conditions

The boundary conditions for the Donnell theory are specified in Sect. 6.6. The way in which we must enhance them follows best from an application of the principle of virtual work. Without providing further elaboration, we advance the boundary conditions here. Figure 9.3 depicts those quantities that may play a role. We again apply the Kirchhoff shear force

$$^*v_x = v_x + \frac{\partial m_{\theta x}}{a\,\partial\theta}, \qquad (9.20)$$

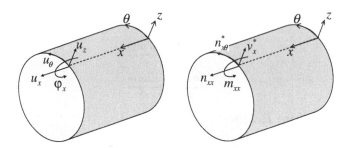

Fig. 9.3 Quantities that occur in boundary conditions

but in the Morley theory also a *generalized membrane shear force* $*n_{x\theta}$ appears:

$$*n_{x\theta} = n_{x\theta} + \frac{2}{a} m_{x\theta}. \tag{9.21}$$

Finally, for full cylindrical shells with a circular edge, we need a definition of the rotation in axial direction:

$$\varphi_x = -\frac{\partial u_z}{\partial x}. \tag{9.22}$$

On the basis of the above definitions, we are able to specify the boundary conditions at the curved edges:

$$
\begin{aligned}
&- \text{ either } u_x \text{ or } n_{xx} \\
&- \text{ either } u_\theta \text{ or } *n_{x\theta} \\
&- \text{ either } u_z \text{ or } *v_x \\
&- \text{ either } \varphi_x \text{ or } m_{xx}
\end{aligned} \tag{9.23}
$$

9.3 Differential Equations for Load p_z

9.3.1 Differential Equations for Displacements

Up to this point, we have introduced no simplifications or assumptions, apart from first-order approximations, choice of load cases and symmetry. To obtain convenient differential equations for displacements, we assume that the parameters describing the material properties and the cross-sectional geometry, i.e., E, v, and a, t respectively, are constant for the whole circular cylindrical shell.

Substitution of the kinematical relation (9.14) into the constitutive relation (9.15) results in what is sometimes referred to as the "elastic law":

$$n_{xx} = D_m \left(\frac{\partial u_x}{\partial x} + v \frac{1}{a} \frac{\partial u_\theta}{\partial \theta} + v \frac{u_z}{a} \right)$$

$$n_{\theta\theta} = D_m \left(v \frac{\partial u_x}{\partial x} + \frac{1}{a} \frac{\partial u_\theta}{\partial \theta} + \frac{u_z}{a} \right) \tag{9.24a}$$

$$n_{x\theta} = D_m \frac{1-v}{2} \left(\frac{1}{a} \frac{\partial u_x}{\partial \theta} + \frac{\partial u_\theta}{\partial x} \right)$$

$$m_{xx} = -D_b \left(\frac{\partial^2 u_z}{\partial x^2} + v \frac{1}{a^2} \frac{\partial^2 u_z}{\partial \theta^2} + v \frac{u_z}{a^2} \right)$$

$$m_{\theta\theta} = -D_b \left(v \frac{\partial^2 u_z}{\partial x^2} + \frac{1}{a^2} \frac{\partial^2 u_z}{\partial \theta^2} + \frac{u_z}{a^2} \right) \tag{9.24b}$$

$$m_{x\theta} = -D_b(1-v) \left(-\frac{1}{a} \frac{\partial u_\theta}{\partial x} + \frac{1}{a} \frac{\partial^2 u_z}{\partial x \partial \theta} \right).$$

Substitution of this elastic law into Eq. (9.18) yields the following three differential equations for the displacements:

$$-\frac{\partial^2 u_x}{\partial x^2} - \frac{1-v}{2} \frac{1}{a^2} \frac{\partial^2 u_x}{\partial \theta^2} - \frac{1+v}{2} \frac{1}{a} \frac{\partial^2 u_\theta}{\partial x \partial \theta} - v \frac{1}{a} \frac{\partial u_z}{\partial x} = 0 \tag{9.25a}$$

$$-\frac{1+v}{2} \frac{1}{a} \frac{\partial^2 u_x}{\partial x \partial \theta} - \frac{1-v}{2} \frac{\partial^2 u_\theta}{\partial x^2} - \frac{1}{a^2} \frac{\partial^2 u_\theta}{\partial \theta^2} - \frac{1}{a^2} \frac{\partial u_z}{\partial \theta}$$
$$-\frac{D_b}{D_m a^2} 2(1-v) \frac{\partial^2 u_\theta}{\partial x^2} + \frac{D_b}{D_m a^2} 2(1-v) \frac{\partial^3 u_z}{\partial x^2 \partial \theta} = 0 \tag{9.25b}$$

$$v \frac{1}{a} \frac{\partial u_x}{\partial x} + \frac{1}{a^2} \frac{\partial u_\theta}{\partial \theta} + \frac{1}{a^2} u_z - \frac{D_b}{D_m a^2} 2(1-v) \frac{\partial^3 u_\theta}{\partial x^2 \partial \theta}$$
$$+ \frac{D_b}{D_m} \left(\frac{\partial^4 u_z}{\partial x^4} + \frac{2}{a^2} \frac{\partial^4 u_z}{\partial x^2 \partial \theta^2} + \frac{1}{a^4} \frac{\partial^4 u_z}{\partial \theta^4} + \frac{2v}{a^2} \frac{\partial^2 u_z}{\partial x^2} + \frac{2}{a^4} \frac{\partial^2 u_z}{\partial \theta^2} + \frac{u_z}{a^4} \right) = \frac{p_z}{D_m}. \tag{9.25c}$$

The three differential equations are symbolically described by

$$\begin{bmatrix} L_{11} & L_{12} & L_{13} \\ L_{21} & L_{22} & L_{23} \\ L_{31} & L_{32} & L_{33} \end{bmatrix} \begin{bmatrix} u_x \\ u_\theta \\ u_z \end{bmatrix} = \frac{1}{D_m} \begin{bmatrix} 0 \\ 0 \\ p_z \end{bmatrix}. \tag{9.26}$$

The operators L_{11} up to and including L_{33} form a differential operator matrix, in which the operators are

$$L_{11} = -\Delta + \frac{1+v}{2}\frac{1}{a^2}\frac{\partial^2}{\partial\theta^2}$$

$$L_{22} = -\Delta + \frac{1+v}{2}\frac{\partial^2}{\partial x^2} - 2k(1-v)\frac{\partial^2}{\partial x^2} \qquad (9.27a)$$

$$L_{33} = \frac{1}{a^2} + ka^2\left(\Delta + \frac{1}{a^2}\right)^2 - 2ka^2(1-v)\frac{\partial^2}{\partial x^2}$$

$$L_{12} = L_{21} = -\frac{1+v}{2}\frac{1}{a}\frac{\partial^2}{\partial x\partial\theta}$$

$$L_{13} = -L_{31} = -v\frac{1}{a}\frac{\partial}{\partial x} \qquad (9.27b)$$

$$L_{23} = -L_{32} = -\frac{1}{a^2}\frac{\partial}{\partial\theta} + 2k(1-v)\frac{\partial^3}{\partial x^2\partial\theta}.$$

Here the Laplacian Δ is defined by Eq. (9.2) and the dimensionless parameter k is introduced, which is defined by

$$k = \frac{D_b}{D_m a^2} = \frac{t^2}{12a^2}. \qquad (9.28)$$

For a thin shell where $t < a$, the parameter k is negligibly small in comparison to unity ($k \ll 1$).

9.3.2 Single Differential Equation

By eliminating u_x from the first and second equation, we obtain the differential equation describing the relation between u_θ and u_z. Equivalently, we eliminate u_θ from the first and second equation to obtain a relation between u_x and u_z. The resulting equations symbolically read

$$(L_{11}L_{22} - L_{21}L_{12})u_\theta + (L_{11}L_{23} - L_{21}L_{13})u_z = 0$$
$$(L_{22}L_{11} - L_{12}L_{21})u_x + (L_{22}L_{13} - L_{12}L_{23})u_z = 0. \qquad (9.29)$$

By substituting these two relations into the third equation, the single differential equation for the displacement u_z is obtained, which symbolically reads

$$[L_{31}(L_{12}L_{23} - L_{22}L_{13}) + L_{32}(L_{21}L_{13} - L_{11}L_{23}) + L_{33}(L_{22}L_{11} - L_{12}L_{21})]u_z$$
$$= \frac{1}{D_m}(L_{22}L_{11} - L_{12}L_{21})p_z. \qquad (9.30)$$

In working out the multiplication of derivatives, we neglect terms with the square of the parameter k in comparison to unity and we account for the dominance of terms that are multiplied by the parameter $a\beta$. In this way, we can set up a simplified and, from a mathematical point of view considerably more elegant, differential equation, practically without loss of accuracy. For the other two equations relating u_x and u_θ to u_z, a similar observation leads again to the neglect of small terms. In this way, we end up with three differential equations of much simpler appearance:

$$\Delta\Delta u_\theta + (2+v)\frac{1}{a^2}\frac{\partial^3 u_z}{\partial x^2 \partial\theta} + \frac{1}{a^4}\frac{\partial^3 u_z}{\partial\theta^3} = 0$$

$$\Delta\Delta u_x + v\frac{1}{a}\frac{\partial^3 u_z}{\partial x^3} - \frac{1}{a^3}\frac{\partial^3 u_z}{\partial x\partial\theta^2} = 0 \qquad (9.31)$$

$$\Delta\Delta\left(\Delta + \frac{1}{a^2}\right)^2 u_z + 4\beta^4\frac{\partial^4 u_z}{\partial x^4} = \frac{1}{D_b}\Delta\Delta p_z$$

The last one is the Morley equation. The parameter β is defined by Eq. (9.3), viz. the same as in the Donnell theory. In comparison with the Donnell equation in Eq. (9.1), the change looks minor. Hereafter, in Chaps. 12 and 13, we will see that the impact is major.

9.4 Homogeneous Solution of the Differential Equation for a Curved Edge

9.4.1 Exact Solution

As stated in Sect. 9.1.4, all quantities for the three load-deformation behaviours can be described by functions of the type of Eqs. (9.7) and (9.8) depending on the axis of symmetry of the quantity under consideration. Hence, we can make the following substitutions for the load and displacements:

$$p_x(x,\theta) = 0; \qquad\qquad u_x(x,\theta) = u_{xn}(x)\cos n\theta$$
$$p_\theta(x,\theta) = 0; \qquad\qquad u_\theta(x,\theta) = u_{\theta n}(x)\sin n\theta \qquad (9.32)$$
$$p_z(x,\theta) = p_{zn}(x)\cos n\theta; \qquad u_z(x,\theta) = u_{zn}(x)\cos n\theta.$$

For derivates with respect to the circumferential coordinate θ and consequently for the Laplacian Δ, we can make substitutions of the form

$$\frac{\partial(x,\theta)}{\partial\theta} = \frac{\partial_n(x)\cos n\theta}{\partial\theta} = -n_n(x)\sin n\theta$$

$$\Delta(x,\theta) = \left(\frac{d^2}{dx^2} - \frac{n^2}{a^2}\right)_n (x)\cos n\theta. \qquad (9.33)$$

Similarly we can do the same for quantities generally described by Eq. (9.8). By substitution of the load and displacement functions given above, the third differential equation in Eq. (9.31) becomes an ordinary differential equation and by omitting the cosine function for the circumferential distribution, the governing differential equation is reduced to

$$\left[\left(\frac{d^2}{dx^2} - \frac{n^2}{a^2}\right)^2 \left(\frac{d^2}{dx^2} - \frac{n^2-1}{a^2}\right)^2 + 4\beta^4 \frac{d^4}{dx^4}\right] u_{zn}(x)$$

$$= \frac{1}{D_b}\left(\frac{d^2}{dx^2} - \frac{n^2}{a^2}\right)^2 p_{zn}(x) \tag{9.34}$$

The homogeneous equation is given by

$$\left[\left(\frac{d^2}{dx^2} - \frac{n^2}{a^2}\right)^2 \left(\frac{d^2}{dx^2} - \frac{n^2-1}{a^2}\right)^2 + 4\beta^4 \frac{d^4}{dx^4}\right] u_{zn}(x) = 0. \tag{9.35}$$

The periodic trial function for $u_z(x, \theta)$ is $u_z(x, \theta) = F_n(x)\cos n\theta$. This function satisfies the continuity and symmetry conditions for a closed circular cylindrical shell. We obtain the solution to the homogeneous equation by similar steps as in Sect. 7.5:

$$\begin{aligned} u_{zn}(x) &= e^{-a_n^1 \beta x}\left[C_1^n \cos\left(b_n^1 \beta x\right) + C_2^n \sin\left(b_n^1 \beta x\right)\right] \\ &+ e^{a_n^1 \beta x}\left[C_3^n \cos\left(b_n^1 \beta x\right) + C_4^n \sin\left(b_n^1 \beta x\right)\right] \\ &+ e^{-a_n^2 \beta x}\left[C_5^n \cos\left(b_n^2 \beta x\right) + C_6^n \sin\left(b_n^2 \beta x\right)\right] \\ &+ e^{a_n^2 \beta x}\left[C_7^n \cos\left(b_n^2 \beta x\right) + C_8^n \sin\left(b_n^2 \beta x\right)\right] \end{aligned} \tag{9.36}$$

where the dimensionless roots a_n^1, a_n^2, b_n^1 and b_n^2 are defined by

$$a_n^1 = \frac{1}{\sqrt{2}}\sqrt{\delta_1 + \gamma + \sqrt{\frac{\omega_1 - 1}{2}}}, \qquad \frac{1}{n} = \frac{1}{\sqrt{2}}\sqrt{\delta_1 - \gamma - \sqrt{\frac{\omega_1 - 1}{2}}}$$

$$a_n^2 = \frac{1}{\sqrt{2}}\sqrt{\delta_2 + \gamma - \sqrt{\frac{\omega_1 - 1}{2}}}, \qquad b_n^2 = \frac{1}{\sqrt{2}}\sqrt{\delta_2 - \gamma + \sqrt{\frac{\omega_1 - 1}{2}}} \tag{9.37}$$

in which

$$\delta_1 = \sqrt{\gamma + \frac{1}{2}(\omega_1 + \omega_2) + 1 + \gamma\sqrt{2(\omega_1 - 1)} + \sqrt{2(\omega_2 + 1)}}, \qquad \omega_1 = \omega + \gamma^2 - \eta^2$$

$$\delta_2 = \sqrt{\gamma + \frac{1}{2}(\omega_1 + \omega_2) + 1 - \gamma\sqrt{2(\omega_1 - 1)} - \sqrt{2(\omega_2 + 1)}}, \qquad \omega_2 = \omega - \gamma^2 + \eta^2$$

$$\tag{9.38}$$

and

$$\omega = \sqrt{1 + 2(\gamma^2 + \eta^2) + (\gamma^2 - \eta^2)^2}$$

$$\gamma = \left(n^2 - \frac{1}{2}\right)/(a\beta)^2, \quad \eta = \sqrt{n(n^2 - 1)}/(a\beta)^2.$$

$$(9.39)$$

The roots in Eq. (9.37) are surplus to requirements and therefore, in the next section, we can make approximations for several load-deformation regimes. The presented solution is a unification of former results by other authors [2, 8].

It deserves attention that the solution for the axisymmetric mode ($n = 0$) and beam mode ($n = 1$) is retained. For these two values of n, the parameter η of Eq. (9.39) is equal to zero and the eight roots are calculated with the reduced parameters

$$a_{n=0,1}^1 = \sqrt{\left(1 + \gamma_{n=0,1}^2\right)^{\frac{1}{2}} + \gamma_{n=0,1}}, \qquad a_{n=0,1}^2 = 0$$

$$b_{n=0,1}^1 = \sqrt{\left(1 + \gamma_{n=0,1}^2\right)^{\frac{1}{2}} - \gamma_{n=0,1}}, \qquad b_{n=0,1}^2 = 0.$$

$$(9.40)$$

Because $a_{n=0,1}^2$ and $b_{n=0,1}^2$ are zero for the cases $n = 0$ and $n = 1$, the last two lines of the expression in Eq. (9.36) reduces to the constant $C_5^n + C_7^n$, which has to be put at zero.

9.4.2 Approximate Solution

To express the dimensionless roots of Eq. (9.37), we have introduced the parameters γ and η of Eq. (9.39). By definition of Eq. (9.3) the parameter $(a\beta)^{-2}$ is a small value for the usual thickness-over-radius-ratio t/a. For the static behaviour of thin shells under the usual loading cases, only the first and lower values of the mode number n are important (say $n = 1, \ldots, 5$) and, therefore, γ and η remain small in comparison to unity. This enables a tremendous reduction of the expressions for the eight roots by expanding these into a series development and then breaking them down after the second term since $\gamma^2 \approx \eta^2 \ll 1$.

For the two lowest modes with $n = 0$ and $n = 1$, parameter η is zero, and we obtain the following approximate expressions:

$$n = 0: \quad a_0^1 = 1 - \frac{1}{4(a\beta)^2}, \quad b_0^1 = 1 + \frac{1}{4(a\beta)^2}$$

$$n = 1: \quad a_1^1 = 1 + \frac{1}{4(a\beta)^2}, \quad b_1^1 = 1 - \frac{1}{4(a\beta)^2}$$

$$a_0^2 = b_0^2 = 0$$

$$a_1^2 = b_1^2 = 0.$$

$$(9.41)$$

The subscripts denote the mode number n.

For $n > 1$, we obtain the following approximate expressions:

$$a_n^1 = 1 + \frac{1}{2}\gamma_n, \qquad a_n^2 = \frac{1}{2}\eta_n\left(1 + \frac{1}{2}\gamma_n\right)$$

$$b_n^1 = 1 - \frac{1}{2}\gamma_n, \qquad b_n^2 = \frac{1}{2}\eta_n\left(1 - \frac{1}{2}\gamma_n\right), \tag{9.42}$$

from which the solution for $n = 0$ and $n = 1$ is still traceable. The parameters γ_n and η_n are identical to the parameters γ and η in Eq. (9.39), but the subscript n is adopted hereafter to indicate the mode numbers $n > 1$. For $n > 1$ and larger values of γ_n and η_n, we may use both the Morley solution and the solution to Donnell's equation. In this domain the differences are minor.

9.5 Influence Length

9.5.1 Axisymmetric Mode

For the axisymmetric mode, we just need to consider in Eq. (9.36) the terms multiplied with C_1 up to and including C_4. The other part of the solution is zero. The terms multiplied with C_1 and C_2 are oscillating functions of the ordinate x that decrease exponentially with increasing x. The terms multiplied with C_3 and C_4 are also damped oscillations but these decrease exponentially with decreasing x. Now the characteristic length l_c and influence length l_i are

$$l_c = \frac{1}{a_0\beta}; \quad l_i = \pi l_c. \tag{9.43}$$

By using the approximate value for a_0 of Eq. (9.41) the characteristic length can be approximated by

$$l_c = \frac{\sqrt{at}}{\sqrt[4]{3(1 - v^2)}} \tag{9.44}$$

in which also $(a\beta)^{-2}$ has been neglected in comparison to unity. This is the same formula that we obtained in Chap. 5 for edge disturbances in axially symmetric shells. The same holds for the influence length:

$$l_i = \pi l_c \approx 2.4\sqrt{at}. \tag{9.45}$$

9.5.2 Beam Mode

For the beam mode we again need to consider in Eq. (9.36) only the terms multiplied with C_1 up to and including C_4. The terms multiplied with C_1 and C_2, are oscillating functions of the ordinate x that decrease exponentially with increasing x. The terms multiplied with C_3 and C_4 are also damped oscillations but these decrease exponentially with decreasing x. The characteristic and influence lengths for this part of the solution are

$$l_c = \frac{1}{a_1 \beta}; \ l_i = \pi l_c. \tag{9.46}$$

By using the approximate value for a_1 of Eq. (9.41) and neglecting $(a\beta)^{-2}$ in comparison to unity, the characteristic length and influence length become

$$l_c \approx \frac{\sqrt{at}}{\sqrt[4]{3(1 - v^2)}}; \ l_i = l_c \approx 2.4\sqrt{at}. \tag{9.47}$$

These are equal to the characteristic and influence length for axisymmetric behaviour.

9.5.3 Self-Balancing Modes

For the self-balancing modes, all eight terms in Eq. (9.36) must be considered. The terms multiplied with the constants C_1^n, C_2^n, C_3^n and C_4^n represent the part of the solution describing an edge disturbance with a short influence length, which is further referred to as the *short-wave solution*. Similarly, the terms multiplied with the constants C_5^n, C_6^n, C_7^n and C_8^n represent the part of the solution describing an edge disturbance with a long influence length, which is further referred to as the *long-wave solution*. The terms multiplied with C_1^n, C_2^n, C_5^n and C_6^n, are oscillating functions of the ordinate x that decrease exponentially with increasing x. The terms multiplied with C_3^n, C_4^n, C_7^n and C_8^n are also damped oscillations but these decrease exponentially with decreasing x. The characteristic and influence lengths for the short-wave solution appear to be equal to the characteristic and influence lengths for the axisymmetric load. We calculate them by

$$l_{c,1} = \frac{1}{a_n^1 \beta}; \qquad l_{i,1} = \pi l_{c,1}. \tag{9.48}$$

By using the approximate value for a_n^1 of Eq. (9.42) and neglecting $(a\beta)^{-2}$ in comparison to unity, this short characteristic and influence length become

$$l_{c,1} \approx \frac{\sqrt{at}}{\sqrt[4]{3(1 - v^2)}}; \ l_{i,1} = l_{c,1} \approx 2.4\sqrt{at}, \quad (9.49)$$

indeed equal to the axisymmetric mode. The characteristic and influence lengths for the long-wave solution are

$$l_{c,2} = \frac{1}{a_n^2 \beta}; \qquad l_{i,2} = \pi l_{c,2}. \quad (9.50)$$

By using the approximate value for a_n^2 of Eq. (9.42) and neglecting $(a\beta)^{-2}$ in comparison to unity, these lengths become

$$l_{c,2} \approx \frac{2}{\eta_n \beta} = \frac{2\sqrt[4]{3(1 - v^2)}}{n\sqrt{n^2 - 1}} \frac{a}{t} \sqrt{at}; \ l_{i,2} \approx \frac{8.1}{n\sqrt{n^2 - 1}} \frac{a}{t} \sqrt{at}. \quad (9.51)$$

These lengths depend on the mode number n. The long influence length describes a far-reaching influence, in the order of a/t times the short influence length $l_{i,1}$. This influence length decreases rapidly with increasing n.

In Chap. 7, the Donnell theory for shallow shells with straight edges, we obtained earlier two different influence lengths. There, the difference was on the order of a factor 2, and the largest one was about equal to the radius a. For disturbances starting from circular edges in the Morley theory, the two influence lengths are far more different. For the lowest self-balancing mode ($n = 2$) the largest length is about $2.3(a/t)\sqrt{at}$, which we also may write as $2.3a\sqrt{a/t}$. The value a/t is of the order 100. Then, the longest influence length is more than 20 times the radius a.

9.6 Displacements and Shell Forces of the Homogeneous Solution for Self-Balancing Modes

Now that we have a homogeneous solution for the displacement u_z, it is necessary to express the other displacements, membrane forces, moments and transverse shear forces as functions of u_z. We can do so for the displacements u_x and u_θ by solving the first two equations of Eq. (9.31):

$$\Delta\Delta u_\theta = -(2 + v)\frac{1}{a^2}\frac{\partial^3 u_z}{\partial x^2 \partial \theta} - \frac{1}{a^4}\frac{\partial^3 u_z}{\partial \theta^3}$$

$$\Delta\Delta u_x = -v\frac{1}{a}\frac{\partial^3 u_z}{\partial x^3} + \frac{1}{a^3}\frac{\partial^3 u_z}{\partial x \partial \theta^2}. \quad (9.52)$$

Furthermore, we write the third equation of Eq. (9.31) as

$$u_z = -\frac{1}{4}\left(\frac{1}{\beta}\right)^4 \int \int \int \int \Delta\Delta\left(\Delta+\frac{1}{a^2}\right)^2 u_z dxdxdxdx. \tag{9.53}$$

By substituting this equation into Eq. (9.52), and omitting two Laplace operators, we obtain the following equations:

$$u_\theta = \frac{1}{4}\left(\frac{1}{\beta}\right)^4 \left[(2+v)\frac{1}{a^2}\frac{\partial^3}{\partial x^2 \partial\theta} + \frac{1}{a^4}\frac{\partial^3}{\partial\theta^3}\right]\int \int \int \int \left(\Delta+\frac{1}{a^2}\right)^2 u_z dxdxdxdx$$

$$u_x = \frac{1}{4}\left(\frac{1}{\beta}\right)^4 \left[v\frac{1}{a}\frac{\partial^3}{\partial x^3} - \frac{1}{a^3}\frac{\partial^3}{\partial x\partial\theta^2}\right]\int \int \int \int \left(\Delta+\frac{1}{a^2}\right)^2 u_z dxdxdxdx. \tag{9.54}$$

By substitution of the displacement functions given above, these become ordinary differential equations, in which we omit the sine function (for u_θ) and the cosine function (for u_x).

The rotation φ_x is

$$\varphi_x = -\frac{\partial u_z}{\partial x}. \tag{9.55}$$

The membrane forces are described by Eq. (9.24a). Upon substitution of Eq. (9.54) the expressions read

$$n_{xx} = -D_b a \int \int \frac{\partial^2\left(\Delta+\frac{1}{a^2}\right)^2 u_z}{a^2\partial\theta^2} dxdx$$

$$n_{\theta\theta} = -D_b a\left(\Delta+\frac{1}{a^2}\right)^2 u_z \tag{9.56}$$

$$n_{x\theta} = D_b a \int \frac{\partial\left(\Delta+\frac{1}{a^2}\right)^2 u_z}{a\partial\theta} dx.$$

The moments are also described by Eq. (9.24b). The moments m_{xx} and $m_{\theta\theta}$ are already expressions of u_z. Upon substitution of the expression for u_θ above, this becomes also the case for moment $m_{x\theta}$:

$$m_{x\theta} = D_b(1-v)\frac{a}{4(a\beta)^4}$$

$$\cdot\left[(2+v)\int \frac{\partial\left(\Delta+\frac{1}{a^2}\right)^2 u_z}{\partial\theta} dx + \frac{1}{a^2}\int \int \int \frac{\partial^3\left(\Delta+\frac{1}{a^2}\right)^2 u_z}{\partial\theta^3} dxdxdx\right] \tag{9.57}$$

$$- D_b(1-v)\frac{1}{a}\frac{\partial^2 u_z}{\partial x\partial\theta}.$$

The transverse shear forces are described by Eq. (9.19). They become

$$
v_x = D_b(1-v)\frac{1}{4(a\beta)^4}
$$

$$
\cdot \left[(2+v)\int \frac{\partial^2 \left(\Delta + \frac{1}{a^2}\right)^2 u_z}{\partial\theta^2}\,dx + \frac{1}{a^2}\int\int\int \frac{\partial^4 \left(\Delta + \frac{1}{a^2}\right)^2 u_z}{\partial\theta^4}\,dxdxdx\right] \tag{9.58a}
$$

$$
+ D_b\left(-\frac{\partial^3 u_z}{\partial x^3} - \frac{1}{a^2}\frac{\partial^3 u_z}{\partial x\partial\theta^2} - v\frac{1}{a^2}\frac{\partial u_z}{\partial x}\right)
$$

$$
v_\theta = D_b(1-v)\frac{a}{4(a\beta)^4}
$$

$$
\cdot \left[(2+v)\frac{\partial \left(\Delta + \frac{1}{a^2}\right)^2 u_z}{\partial\theta} + \frac{1}{a^2}\int\int \frac{\partial^3 \left(\Delta + \frac{1}{a^2}\right)^2 u_z}{\partial\theta^3}\,dxdx\right] \tag{9.58b}
$$

$$
+ D_b\frac{1}{a}\left(-\frac{\partial^3 u_z}{\partial x^2\partial\theta} - \frac{1}{a^2}\frac{\partial^3 u_z}{\partial\theta^3} - \frac{1}{a^2}\frac{\partial u_z}{\partial\theta}\right).
$$

The Kirchhoff shear force $^*v_x$ of Eq. (9.20) becomes

$$
{}^*V_x = D_b(1-v)\frac{1}{2(a\beta)^4}
$$

$$
\cdot \left[(2+v)\int \frac{\partial^2 \left(\Delta + \frac{1}{a^2}\right)^2 u_z}{\partial\theta^2}\,dx + \frac{1}{a^2}\int\int\int \frac{\partial^4 \left(\Delta + \frac{1}{a^2}\right)^2 u_z}{\partial\theta^4}\,dxdxdx\right] \tag{9.59}
$$

$$
+ D_b\left(-\frac{\partial^3 u_z}{\partial x^3} - \frac{2-v}{a^2}\frac{\partial^3 u_z}{\partial x\partial\theta^2} - v\frac{1}{a^2}\frac{\partial u_z}{\partial x}\right).
$$

Finally, the generalized membrane shear force of Eq. (9.21) becomes

$$
n_{x\theta}^* = D_b\left(1 + (1-v)(2+v)\frac{1}{2(a\beta)^4}\right)\int \frac{\partial \left(\Delta + \frac{1}{a^2}\right)^2 u_z}{\partial\theta}\,dx
$$

$$
+ D_b(1-v)\frac{1}{2a^2(a\beta)^4}\int\int\int \frac{\partial^3 \left(\Delta + \frac{1}{a^2}\right)^2 u_z}{\partial\theta^3}\,dxdxdx \tag{9.60}
$$

$$
- D_b(1-v)\frac{1}{a}\frac{\partial^2 u_z}{\partial x\partial\theta}.
$$

By working out the derivatives and integrals, we obtain all displacements and shell forces as functions of the coordinates x and θ multiplied by eight constants of the homogeneous solution. The resulting expressions are presented in Table 9.1.

9.6.1 Comparison with Donnell Solution

For purposes of comparison in Chap. 12 on chimneys, we want to know the differences with the Donnell solution. For the Donnell equation in Eq. (9.1) and by adopting the expressions for shell forces and displacements given in Sect. 7.4, those resulting expressions can be similarly obtained. The main change is that the operator $\Delta + 1/a^2$ is replaced by Δ only. The result is Table 9.2. If we apply the same approximation as done in Sect. 9.4.2 for Morley, the table changes into Table 9.3. As stated at the end of Sect. 9.3, the change in differential equation looks minor, but substantial differences occur between Table 9.1 for the Morley equation and Table 9.3 for the Donnell equation.

9.7 Inhomogeneous Solution for Self-Balancing Modes

Restricting us to constant or linear loads p_z, we can reduce the inhomogeneous equation of Eq. (9.34) to

$$\frac{n^4}{a^4}\left(\frac{n^2-1}{a^2}\right)^2 u_{zn}(x) = \frac{1}{D_b}\frac{n^4}{a^4}p_{zn}(x). \tag{9.61}$$

The inhomogeneous solution is

$$u_{zn}(x) = \frac{1}{D_b}\left(\frac{a^2}{n^2-1}\right)^2 p_{zn}(x). \tag{9.62}$$

By substituting this result into the first two expressions of Eq. (9.31), we can obtain the inhomogeneous solution for the circumferential displacement u_θ and the axial displacement u_x. If we omit the second and higher derivatives with respect to x, these differential equations become

$$\frac{1}{a^4}\frac{\partial^4 u_\theta}{\partial\theta^4} + \frac{1}{a^4}\frac{\partial^3 u_z}{\partial\theta^3} = 0$$
$$\frac{1}{a^4}\frac{\partial^4 u_x}{\partial\theta^4} - \frac{1}{a^3}\frac{\partial^3 u_z}{\partial x\partial\theta^2} = 0. \tag{9.63}$$

We can rewrite these equations by substituting the displacement and load functions given above. If we omit the cosine and sine terms, the equations become

$$u_{\theta n}(x) = -\frac{1}{n}u_{zn}(x)$$
$$u_{xn}(x) = -\frac{a}{n^2}\frac{du_{zn}(x)}{dx} \tag{9.64}$$

into which the solution of Eq. (9.62) can be substituted.

Table 9.1 The expressions (based on Eq. (9.42)) for the quantities of the homogeneous solution for the circular edge

Multiplicator	$e^{-[(1+\frac{1}{2}\gamma)\beta x]}\cos[(1-\frac{1}{2}\gamma)\beta x]$	$e^{-[(1+\frac{1}{2}\gamma)\beta x]}\sin[(1-\frac{1}{2}\gamma)\beta x]$	$e^{[(1+\frac{1}{2}\gamma)\beta x]}\cos[(1-\frac{1}{2}\gamma)\beta x]$	$e^{[(1+\frac{1}{2}\gamma)\beta x]}\sin[(1-\frac{1}{2}\gamma)\beta x]$
u_z	$\cos(n\theta)$	C_1	C_2	C_3	C_4

	Multiplicator	col1	col2	col3	col4
u_z	$\cos(n\theta)$	C_1	C_2	C_3	C_4
φ_x	$\frac{\beta}{2}\cos(n\theta)$	$(2+\gamma)C_1-(2-\gamma)C_2$	$(2-\gamma)C_1+(2+\gamma)C_2$	$-(2+\gamma)C_3-(2-\gamma)C_4$	$(2-\gamma)C_3+(2+\gamma)C_4$
u_x	$\frac{1}{4a\beta}\cos(n\theta)$	$u_x^{11}C_1-u_x^{12}C_2$	$u_x^{12}C_1+u_x^{11}C_2$	$-u_x^{11}C_3-u_x^{12}C_4$	$u_x^{12}C_3-u_x^{11}C_4$
u_θ	$\frac{1}{2n}\sin(n\theta)$	$u_\theta^{11}C_1-u_\theta^{12}C_2$	$u_\theta^{12}C_1+u_\theta^{11}C_2$	$u_\theta^{11}C_3+u_\theta^{12}C_4$	$-u_\theta^{12}C_3+u_\theta^{11}C_4$
m_{xx}	$D_b\beta^2\cos(n\theta)$	$-\left(2\gamma-\nu\frac{n^2-1}{a^2\beta^2}\right)C_1+2C_2$	$-2C_1-\left(2\gamma-\nu\frac{n^2-1}{a^2\beta^2}\right)C_2$	$-\left(2\gamma-\nu\frac{n^2-1}{a^2\beta^2}\right)C_3-2C_4$	$2C_3-\left(2\gamma-\nu\frac{n^2-1}{a^2\beta^2}\right)C_4$
$m_{\theta\theta}$	$D_b\beta^2\cos(n\theta)$	$\left(\frac{n^2-1}{a^2\beta^2}-2\nu\gamma\right)C_1+2\nu C_2$	$-2\nu C_1+\left(\frac{n^2-1}{a^2\beta^2}-2\nu\gamma\right)C_2$	$\left(\frac{n^2-1}{a^2\beta^2}-2\nu\gamma\right)C_3-2\nu C_4$	$2\nu C_3+\left(\frac{n^2-1}{a^2\beta^2}-2\nu\gamma\right)C_4$
$m_{x\theta}$	$D_b\beta\frac{1-\nu n}{2}\frac{n}{a}\sin(n\theta)$	$m_{x\theta}^{11}C_1-m_{x\theta}^{12}C_2$	$m_{x\theta}^{12}C_1+m_{x\theta}^{11}C_2$	$-m_{x\theta}^{11}C_3-m_{x\theta}^{12}C_4$	$m_{x\theta}^{12}C_3-m_{x\theta}^{11}C_4$
v_x^*	$D_b\beta^3\sin(n\theta)$	$v_x^{*11}C_1-v_x^{*12}C_2$	$v_x^{*12}C_1+v_x^{*11}C_2$	$-v_x^{*11}C_3-v_x^{*12}C_4$	$v_x^{*12}C_3-v_x^{*11}C_4$
v_x	$D_b\beta^3\cos(n\theta)$	$v_x^{11}C_1-v_x^{12}C_2$	$v_x^{12}C_1+v_x^{11}C_2$	$v_x^{11}C_3-v_x^{12}C_4$	$v_x^{12}C_3-v_x^{11}C_4$
v_θ	$D_b\beta^2\frac{n}{a}\sin(n\theta)$	$\left(\frac{n^2}{a^2\beta^2}+(2+\nu)\frac{1-\nu}{a^2\beta^2}\right)C_1-2C_2$	$2C_1+\left(\frac{n^2}{a^2\beta^2}+(2+\nu)\frac{1-\nu}{a^2\beta^2}\right)C_2$	$\left(\frac{n^2}{a^2\beta^2}+(2+\nu)\frac{1-\nu}{a^2\beta^2}\right)C_3+2C_4$	$-2C_3+\left(\frac{n^2}{a^2\beta^2}+(2+\nu)\frac{1-\nu}{a^2\beta^2}\right)C_4$
n_{xx}	$\frac{Et}{4a^2\beta^2}\left(\frac{n}{a}\right)^2 a\cos(n\theta)$	$\frac{1}{a^2\beta^2}C_1-2C_2$	$2C_1+\frac{1}{a^2\beta^2}C_2$	$\frac{1}{a^2\beta^2}C_3+2C_4$	$-2C_3+\frac{1}{a^2\beta^2}C_4$
$n_{\theta\theta}$	$\frac{Et}{a}\cos(n\theta)$	$C_1+\frac{n^2}{a^2\beta^2}C_2$	$-\frac{n^2}{a^2\beta^2}C_1+C_2$	$C_3-\frac{n^2}{a^2\beta^2}C_4$	$\frac{n^2}{a^2\beta^2}C_3+C_4$
$n_{x\theta}$	$\frac{Et}{4a\beta}\frac{n}{a}\sin(n\theta)$	$n_{x\theta}^{11}C_1-n_{x\theta}^{12}C_2$	$n_{x\theta}^{12}C_1+n_{x\theta}^{11}C_2$	$-n_{x\theta}^{11}C_3-n_{x\theta}^{12}C_4$	$n_{x\theta}^{12}C_3-n_{x\theta}^{11}C_4$

(continued)

Table 9.1 (continued)

Multiplicator	$e^{-[\eta(1+\frac{1}{2}\gamma)\frac{\beta}{2}x]}\cos\left[\eta(1-\frac{1}{2}\gamma)\frac{\beta}{2}x\right]$	$e^{-[\eta(1+\frac{1}{2}\gamma)\frac{\beta}{2}x]}\sin\left[\eta(1-\frac{1}{2}\gamma)\frac{\beta}{2}x\right]$	$e^{[\eta(1+\frac{1}{2}\gamma)\frac{\beta}{2}x]}\cos\left[\eta(1-\frac{1}{2}\gamma)\frac{\beta}{2}x\right]$	$e^{[\eta(1+\frac{1}{2}\gamma)\frac{\beta}{2}x]}\sin\left[\eta(1-\frac{1}{2}\gamma)\frac{\beta}{2}x\right]$
u_z $\cos(n\theta)$	C_5	C_6	C_7	C_8
φ_x $\frac{\beta}{2}\cos(n\theta)$	$\eta(1+\frac{1}{2}\gamma)C_5 - \eta(1-\frac{1}{2}\gamma)C_6$	$\eta(1-\frac{1}{2}\gamma)C_5 + \eta(1+\frac{1}{2}\gamma)C_6$	$-\eta(1+\frac{1}{2}\gamma)C_7 - (1-\frac{1}{2}\gamma)C_8$	$(1-\frac{1}{2}\gamma)C_7 - (1+\frac{1}{2}\gamma)C_8$
u_x $\frac{1}{4a\beta}\cos(n\theta)$	$u_x^{21}C_5 - u_x^{22}C_6$	$u_x^{22}C_5 + u_x^{21}C_6$	$-u_x^{21}C_7 - u_x^{22}C_8$	$u_x^{22}C_7 - u_x^{21}C_8$
u_θ $\frac{1}{2n}\sin(n\theta)$	$u_\theta^{21}C_5 - u_\theta^{22}C_6$	$u_\theta^{22}C_5 + u_\theta^{21}C_6$	$u_\theta^{21}C_7 + u_\theta^{22}C_8$	$-u_\theta^{22}C_7 + u_\theta^{21}C_8$
m_{xx} $D_b\beta^2\cos(n\theta)$	$v\frac{n^2-1}{a^2\beta^2}C_5 + \frac{1}{2}\eta^2 C_6$	$-\frac{1}{2}\eta^2 C_5 + v\frac{n^2-1}{a^2\beta^2}C_6$	$v\frac{n^2-1}{a^2\beta^2}C_7 - \frac{1}{2}\eta^2 C_8$	$\frac{1}{2}\eta^2 C_7 + v\frac{n^2-1}{a^2\beta^2}C_8$
$m_{\theta\theta}$ $D_b\beta^2\cos(n\theta)$	$\frac{n^2-1}{a^2\beta^2}C_5$	$\frac{n^2-1}{a^2\beta^2}C_6$	$\frac{n^2-1}{a^2\beta^2}C_7$	$\frac{n^2-1}{a^2\beta^2}C_8$
$m_{x\theta}$ $D_b\beta\frac{1-vn}{2}\frac{n}{a}\sin(n\theta)$	$m_{x\theta}^{21}C_5 - m_{x\theta}^{22}C_6$	$m_{x\theta}^{22}C_5 + m_{x\theta}^{21}C_6$	$-m_{x\theta}^{21}C_7 - m_{x\theta}^{22}C_8$	$m_{x\theta}^{22}C_7 - m_{x\theta}^{21}C_8$
v_x^* $D_b\beta^3\sin(n\theta)$	$v_x^{*21}C_5 - v_x^{*22}C_6$	$v_x^{*22}C_5 + v_x^{*21}C_6$	$v_x^{*21}C_7 - v_x^{*22}C_8$	$v_x^{*22}C_7 - v_x^{*21}C_8$
v_x $D_b\beta^3\cos(n\theta)$	$v_x^{21}C_5 - v_x^{22}C_6$	$v_x^{22}C_5 + v_x^{21}C_6$	$-v_x^{21}C_7 - v_x^{22}C_8$	$v_x^{22}C_7 - v_x^{21}C_8$
v_θ $D_b\beta^2\frac{n}{a}\sin(n\theta)$	$-\frac{n^2-1}{a^2\beta^2}C_5$	$-\frac{n^2-1}{a^2\beta^2}C_6$	$-\frac{n^2-1}{a^2\beta^2}C_7$	$-\frac{n^2-1}{a^2\beta^2}C_8$
n_{xx} $\frac{Et}{4a^2\beta^2}a\cos(n\theta)$	$-\frac{n^2-1}{n^2}\frac{1}{a^2\beta^2}C_5 - 2\frac{n^2-1}{n^2}C_6$	$-2\frac{n^2-1}{n^2}C_5 - \frac{n^2-1}{n^2}\frac{1}{a^2\beta^2}C_6$	$\frac{n^2-1}{n^2}\frac{1}{a^2\beta^2}C_7 - 2\frac{n^2-1}{n^2}C_8$	$2\frac{n^2-1}{n^2}C_7 - \frac{n^2-1}{n^2}\frac{1}{a^2\beta^2}C_8$
$n_{\theta\theta}$ $\frac{Et}{a}\cos(n\theta)$	$-\frac{1}{4}\left(\frac{n^2-1}{a^2\beta^2}\right)^2 C_5$	$-\frac{1}{4}\left(\frac{n^2-1}{a^2\beta^2}\right)^2 C_6$	$-\frac{1}{4}\left(\frac{n^2-1}{a^2\beta^2}\right)^2 C_7$	$-\frac{1}{4}\left(\frac{n^2-1}{a^2\beta^2}\right)^2 C_8$
$n_{x\theta}$ $\frac{Et}{4a\beta}\frac{n}{a}\sin(n\theta)$	$n_{x\theta}^{21}C_5 - n_{x\theta}^{22}C_6$	$n_{x\theta}^{22}C_5 + n_{x\theta}^{21}C_6$	$-n_{x\theta}^{21}C_7 - n_{x\theta}^{22}C_8$	$n_{x\theta}^{22}C_7 - n_{x\theta}^{21}C_8$

(continued)

$$u_x^{11} = -\left(\frac{n^2}{a^2\beta^2}\left(1 - \frac{1}{2}\frac{n^2-\frac{3}{2}}{a^2\beta^2}\right) - 2\nu\left(1 - \frac{1}{2}\frac{n^2-\frac{1}{2}}{a^2\beta^2}\right)\right)$$

$$u_x^{12} = -\left(\frac{n^2}{a^2\beta^2}\left(1 + \frac{1}{2}\frac{n^2-\frac{3}{2}}{a^2\beta^2}\right) + 2\nu\left(1 + \frac{1}{2}\frac{n^2-\frac{1}{2}}{a^2\beta^2}\right)\right)$$

$$u_x^{21} = \frac{n\sqrt{n^2-1}}{a^2\beta^2}\left(2\left(1 - \frac{1}{2}\frac{n^2-\frac{3}{2}}{a^2\beta^2}\right) - \nu\frac{n^2-1}{a^2\beta^2}\right)$$

$$u_x^{22} = \frac{n\sqrt{n^2-1}}{n^2}\left(2\left(1 + \frac{1}{2}\frac{n^2-\frac{3}{2}}{a^2\beta^2}\right) + \nu\frac{n^2-1}{a^2\beta^2}\right)$$

$$m_{x\theta}^{11} = -\left(2 + \frac{n^2+\frac{3}{2}+\nu}{a^2\beta^2}\right)$$

$$m_{x\theta}^{12} = -\left(2 - \frac{n^2+\frac{3}{2}+\nu}{a^2\beta^2}\right)$$

$$m_{x\theta}^{21} = \frac{n\sqrt{n^2-1}}{a^2\beta^2}\frac{n^2-1}{n^2}\left(1 + \frac{1}{2}\frac{n^2-\frac{1}{2}-\nu}{a^2\beta^2}\right)$$

$$m_{x\theta}^{22} = \frac{n\sqrt{n^2-1}}{a^2\beta^2}\frac{n^2-1}{n^2}\left(1 - \frac{1}{2}\frac{n^2-\frac{1}{2}-\nu}{a^2\beta^2}\right)$$

$$v_x^{*11} = -\left(2 - \frac{n^2-\frac{3}{2}+\nu(n^2+1)}{a^2\beta^2}\right)$$

$$v_x^{*12} = -\left(2 + \frac{n^2-\frac{3}{2}+\nu(n^2+1)}{a^2\beta^2}\right)$$

$$v_x^{21} = \frac{2-\nu}{2}\frac{n^2-1}{a^2\beta^2}\frac{n\sqrt{n^2-1}}{a^2\beta^2}$$

$$v_x^{22} = \frac{2-\nu}{2}\frac{n^2-1}{a^2\beta^2}\frac{n\sqrt{n^2-1}}{a^2\beta^2}$$

$$\eta = \frac{n\sqrt{n^2-1}}{a^2\beta^2}$$

$$\gamma = \frac{n^2-\frac{1}{2}}{a^2\beta^2}$$

$$u_\theta^{11} = -\frac{1}{2}\left(\frac{n}{a\beta}\right)^2\frac{n^2-2-\nu}{a^2\beta^2}\left(\frac{n}{a\beta}\right)^2$$

$$u_\theta^{12} = -(2+\nu)\left(\frac{n}{a\beta}\right)^2$$

$$u_\theta^{21} = -2$$

$$u_\theta^{22} = \nu\frac{n^2-1}{a^2\beta^2}$$

$$v_x^{11} = -\left(2 - \frac{2n^2-\frac{3}{2}+\nu}{a^2\beta^2}\right)$$

$$v_x^{12} = \left(2 + \frac{2n^2-\frac{1}{2}+\nu}{a^2\beta^2}\right)$$

$$v_x^{21} = \frac{n^2-1}{a^2\beta^2}\frac{n\sqrt{n^2-1}}{2a^2\beta^2}$$

$$v_x^{22} = \frac{n^2-1}{a^2\beta^2}\frac{n\sqrt{n^2-1}}{2a^2\beta^2}$$

$$n_{x\theta}^{11} = -2\left(1 - \frac{1}{2}\frac{n^2+\frac{1}{2}}{a^2\beta^2}\right)$$

$$n_{x\theta}^{12} = 2\left(1 + \frac{1}{2}\frac{n^2+\frac{1}{2}}{a^2\beta^2}\right)$$

$$n_{x\theta}^{21} = \frac{n^2-1}{n^2}\frac{n\sqrt{n^2-1}}{a^2\beta^2}\left(1 - \frac{1}{2}\frac{n^2+\frac{1}{2}}{a^2\beta^2}\right)$$

$$n_{x\theta}^{22} = -\frac{n^2-1}{n^2}\frac{n\sqrt{n^2-1}}{a^2\beta^2}\left(1 + \frac{1}{2}\frac{n^2+\frac{1}{2}}{a^2\beta^2}\right)$$

Table 9.2 The exact expressions (derived with the Donnell equation (9.1)) for the quantities of the homogeneous solution for the circular edge (note: $\gamma = n^2/(\alpha\beta)^2$)

	Multiplicator	$e^{-k_1}\cdot c_1$	$e^{-k_1}\cdot s_1$	$e^{k_1}\cdot c_1$	$e^{k_1}\cdot s_1$
u_z	$\cos(n\theta)$	C_1	C_2	C_3	C_4
φ_x	$\frac{\beta}{2}\cos(n\theta)$	$-(\sigma+1)C_1+\left(1+\frac{1}{\sigma}\right)C_2$	$-\left(1+\frac{1}{\sigma}\right)C_1-(\sigma+1)C_2$	$(\sigma+1)C_3+\left(1+\frac{1}{\sigma}\right)BC_4$	$-\left(1+\frac{1}{\sigma}\right)C_3+(\sigma+1)C_4$
u_x	$\frac{1}{4\alpha\beta}\cos(n\theta)$	$-\left(\left(1-\frac{1}{\sigma}\right)-\nu\left(1+\frac{1}{\sigma}\right)\right)C_1$ $+\left((\sigma-1)+\nu(\sigma+1)\right)C_2$	$-((\sigma-1)+\nu(\sigma+1))C_1$ $-\left(\left(1-\frac{1}{\sigma}\right)-\nu\left(1+\frac{1}{\sigma}\right)\right)C_2$	$\left(\left(1-\frac{1}{\sigma}\right)-\nu\left(1+\frac{1}{\sigma}\right)\right)C_3$ $+\left((\sigma-1)+\nu(\sigma+1)\right)C_4$	$-((\sigma-1)+\nu(\sigma+1))C_3$ $+\left(\left(1-\frac{1}{\sigma}\right)-\nu\left(1+\frac{1}{\sigma}\right)\right)C_4$
u_θ	$\frac{1}{4n}\sin(n\theta)$	$(E_1-4)C_1+$ $(D_1+2\nu\gamma)C_2$	$-(D_1+2\nu\gamma)C_1$ $+(E_1-4)C_2$	$(E_1-4)C_3-$ $(D_1+2\nu\gamma)C_4$	$(D_1+2\nu\gamma)C_3$ $+(E_1-4)C_4$
m_{xx}	$D_b\frac{\beta^2}{2}\cos(n\theta)$	$-(D_1-2\nu\gamma)C_1+E_1C_2$	$-E_1C_1-(D_1-2\nu\gamma)C_2$	$-(D_1-2\nu\gamma)C_3-E_1C_4$	$E_1C_3-(D_1-2\nu\gamma)C_4$
$m_{\theta\theta}$	$D_b\frac{\beta^2}{2}\cos(n\theta)$	$(2\gamma-\nu D_1)C_1+\nu E_1C_2$	$-\nu E_1C_1+(2\gamma-\nu D_1)C_2$	$(2\gamma-\nu D_1)C_3-\nu E_1C_4$	$\nu E_1C_3+(2\gamma-\nu D_1)C_4$
$m_{x\theta}$	$D_b\beta\frac{1-\nu}{2}\frac{n}{a}\sin(n\theta)$	$-(\sigma+1)C_1+\left(1+\frac{1}{\sigma}\right)C_2$	$-\left(1+\frac{1}{\sigma}\right)C_1-(\sigma+1)C_2$	$(\sigma+1)C_3+\left(1+\frac{1}{\sigma}\right)C_4$	$-(1+\frac{1}{\sigma})C_3+(\sigma+1)C_4$
v_x	$D_b\frac{\beta^3}{4}\cos(n\theta)$	$(2D_1-2E_1)C_1$ $-(2D_1+2E_1)C_2$	$(2D_1+2E_1)C_1$ $+(2D_1-2E_1)C_2$	$(2D_1-2E_1)C_3$ $-(2D_1+2E_1)C_4$	$(2D_1+2E_1)C_3$ $-(2D_1-2E_1)C_4$
v_θ	$D_b\beta^2\frac{n}{a}\sin(n\theta)$	$\frac{1}{2}(\sigma-\frac{1}{\sigma})C_1-\frac{1}{2}E_1C_2$	$\frac{1}{2}E_1BC_1+\frac{1}{2}\left(\sigma-\frac{1}{\sigma}\right)BC_2$	$\frac{1}{2}\left(\sigma-\frac{1}{\sigma}\right)C_3+\frac{1}{2}E_1C_4$	$-\frac{1}{2}E_1C_3+\frac{1}{2}\left(\sigma-\frac{1}{\sigma}\right)C_4$
n_{xx}	$\frac{Et}{4\alpha^2\beta^2}\left(\frac{n}{a}\right)^2 a\cos(n\theta)$	$-C_2$	C_1	C_4	$-C_3$
$n_{\theta\theta}$	$\frac{Et}{4a}\cos(n\theta)$	$E_1C_1+D_1C_2$	$-D_1C_1+E_1C_2$	$E_1C_3-D_1C_4$	$D_1C_3+E_1C_4$
$n_{x\theta}$	$\frac{Et}{4a\beta}\frac{n}{a}\sin(n\theta)$	$-(1+\frac{1}{\sigma})C_1-(\sigma+1)C_2$	$(\sigma+1)C_1-(1+\frac{1}{\sigma})C_2$	$(1+\frac{1}{\sigma})C_3-(\sigma+1)C_4$	$(\sigma+1)C_3+(1+\frac{1}{\sigma})C_1$

(continued)

Table 9.2 (continued)

	Multiplicator	$e^{-k_2}\cdot c_2$	$e^{-k_2}\cdot s_2$	$e^{k_2}\cdot c_2$	$e^{k_2}\cdot s_2$
u_z	$\cos(n\theta)$	C_5	C_6	C_7	C_8
φ_x	$\dfrac{\beta}{2}\cos(n\theta)$	$-(\sigma-1)C_5+(1-\tfrac{1}{\sigma})C_6$	$-(1-\tfrac{1}{\sigma})C_5-(\sigma-1)C_6$	$(\sigma-1)C_7+(1-\tfrac{1}{\sigma})C_8$	$-(1-\tfrac{1}{\sigma})C_7+(\sigma-1)C_8$
u_x	$\dfrac{1}{4a\beta}\cos(n\theta)$	$\left(\left(1+\tfrac{1}{\sigma}\right)-\nu\left(1-\tfrac{1}{\sigma}\right)\right)C_5$ $-((\sigma+1)+\nu(\sigma-1))C_6$	$((\sigma+1)+\nu(\sigma-1))C_5$ $+\left(\left(1+\tfrac{1}{\sigma}\right)-\nu\left(1-\tfrac{1}{\sigma}\right)\right)C_6$	$-\left(\left(1+\tfrac{1}{\sigma}\right)-\nu\left(1-\tfrac{1}{\sigma}\right)\right)C_7$ $-((\sigma+1)+\nu(\sigma-1))C_8$	$((\sigma+1)+\nu(\sigma-1))C_7$ $-\left(\left(1+\tfrac{1}{\sigma}\right)-\nu\left(1-\tfrac{1}{\sigma}\right)\right)C_8$
u_θ	$\dfrac{1}{4n}\sin(n\theta)$	$(E_2-4)C_5-$ $(D_2+2\nu\gamma)C_6$	$(D_2+2\nu\gamma)C_5$ $+(E_2-4)C_6$	$(E_2-4)C_7+$ $(D_2+2\nu\gamma)C_8$	$-(D_2+2\nu\gamma)C_7$ $+(E_2-4)C_8$
m_{xx}	$D_b\dfrac{\beta^2}{2}\cos(n\theta)$	$-(D_2-2\nu\gamma)C_5-E_2C_6$	$E_2C_5-(D_2-2\nu\gamma)C_6$	$-(D_2-2\nu\gamma)C_7+E_2C_8$	$-E_2C_7-(D_2-2\nu\gamma)C_8$
$m_{\theta\theta}$	$D_b\dfrac{\beta^2}{2}\cos(n\theta)$	$(2\gamma-\nu D_2)C_5-\nu E_2C_6$	$\nu E_2C_5+(2\gamma-\nu D_2)C_6$	$(2\gamma-\nu D_2)C_7+\nu E_2C_8$	$-\nu E_2C_7+(2\gamma-\nu D_2)C_8$
$m_{x\theta}$	$D_b\beta\dfrac{1-\nu n}{2}\dfrac{1}{a}\sin(n\theta)$	$-(\sigma-1)C_5+(1-\tfrac{1}{\sigma})C_6$	$-(1-\tfrac{1}{\sigma})C_5-(\sigma-1)C_6$	$(\sigma-1)C_7+(1-\tfrac{1}{\sigma})C_8$	$-(1-\tfrac{1}{\sigma})C_7+(\sigma-1)C_8$
v_x	$D_b\dfrac{\beta^3}{4}\cos(n\theta)$	$-(2D_2-2E_2)C_5$ $-(2D_2+2E_2)C_6$	$(2D_2+2E_2)C_5$ $-(2D_2-2E_2)C_6$	$(2D_2-2E_2)C_7$ $-(2D_2+2E_2)C_8$	$(2D_2+2E_2)C_7$ $+(2D_2-2E_2)C_8$
v_θ	$D_b\beta^2\dfrac{n}{a}\sin(n\theta)$	$-\tfrac{1}{2}(\sigma-\tfrac{1}{\sigma})C_5+\tfrac{1}{2}E_2C_6$	$-\tfrac{1}{2}E_2C_5-\tfrac{1}{2}(\sigma-\tfrac{1}{\sigma})C_6$	$-\tfrac{1}{2}(\sigma-\tfrac{1}{\sigma})C_7-\tfrac{1}{2}E_2C_8$	$\tfrac{1}{2}E_2C_7-\tfrac{1}{2}(\sigma-\tfrac{1}{\sigma})C_8$
n_{xx}	$\dfrac{Et}{4a^2\beta^2}\left(\dfrac{n}{a}\right)^2 a\cos(n\theta)$	C_6	$-C_5$	$-C_8$	C_7
$n_{\theta\theta}$	$\dfrac{Et}{4a}\cos(n\theta)$	$E_2C_5-D_2C_6$	$D_2C_5+E_2C_6$	$E_2C_7+D_2C_8$	$-D_2C_7+E_2C_8$
$n_{x\theta}$	$\dfrac{Et}{4a\beta}\dfrac{n}{a}\sin(n\theta)$	$(1-\tfrac{1}{\sigma})C_5+(\sigma-1)C_6$	$-(\sigma-1)C_5+(1-\tfrac{1}{\sigma})C_6$	$-(1-\tfrac{1}{\sigma})C_7+(\sigma-1)C_8$	$-(\sigma-1)C_7-(1-\tfrac{1}{\sigma})C_8$

(continued)

$$\gamma = \frac{n^2}{a^2\beta^2} = \frac{1}{4}\left(\sigma^2 - \frac{1}{\sigma^2}\right)$$

$$\sigma = \sqrt{4\gamma^2+1} + 2\gamma$$

$$\frac{1}{\sigma} = \sqrt{4\gamma^2+1} - 2\gamma$$

$$D_1 = \frac{\sigma^2}{2} + \sigma - 1 - \frac{1}{2\sigma^2} = \sigma + 2\gamma - \frac{1}{\sigma}$$

$$E_1 = \sigma + 2 + \frac{1}{\sigma}$$

$$D_2 = \frac{\sigma^2}{2} - \sigma + 1 - \frac{1}{2\sigma^2} = -\sigma + 2\gamma + \frac{1}{\sigma}$$

$$E_2 = -\sigma + 2 - \frac{1}{\sigma}$$

$$k_1 = (\sigma+1)\frac{\beta}{2}x$$

$$c_1 = \cos\left(1 + \frac{1}{\sigma}\right)\frac{\beta}{2}x$$

$$s_1 = \sin\left(1 + \frac{1}{\sigma}\right)\frac{\beta}{2}x$$

$$k_2 = (\sigma-1)\frac{\beta}{2}x$$

$$c_2 = \cos\left(1 - \frac{1}{\sigma}\right)\frac{\beta}{2}x$$

$$s_2 = \sin\left(1 - \frac{1}{\sigma}\right)\frac{\beta}{2}x$$

$$u_x^{11} = -\left(\frac{n^2}{a^2\beta^2}\left(1 - \frac{1}{2}\frac{n^2-\frac{3}{2}}{a^2\beta^2}\right) - 2\nu\left(1 - \frac{1}{2}\frac{n^2-\frac{1}{2}}{a^2\beta^2}\right)\right)$$

$$u_x^{12} = -\left(\frac{n^2}{a^2\beta^2}\left(1 + \frac{1}{2}\frac{n^2-\frac{3}{2}}{a^2\beta^2}\right) + 2\nu\left(1 + \frac{1}{2}\frac{n^2-\frac{1}{2}}{a^2\beta^2}\right)\right)$$

$$u_x^{21} = \frac{n\sqrt{n^2-1}}{n^2}\left(2\left(1 - \frac{1}{2}\frac{n^2-\frac{3}{2}}{a^2\beta^2}\right) - \nu\frac{n^2-1}{a^2\beta^2}\right)$$

$$u_x^{22} = \frac{n\sqrt{n^2-1}}{n^2}\left(2\left(1 + \frac{1}{2}\frac{n^2-\frac{3}{2}}{a^2\beta^2}\right) + \nu\frac{n^2-1}{a^2\beta^2}\right)$$

$$u_\theta^{11} = \frac{1}{2}\left(\frac{n}{a\beta}\right)^2\frac{n^2-2-\nu}{a^2\beta^2}\left(\frac{n}{a\beta}\right)^2$$

$$u_\theta^{12} = -(2+\nu)\left(\frac{n}{a\beta}\right)^2$$

$$u_\theta^{21} = -2$$

$$u_\theta^{22} = \nu\frac{n^2-1}{a^2\beta^2}$$

$$m_{x\theta}^{11} = -\left(2 + \frac{n^2+\frac{3}{2}+\nu}{a^2\beta^2}\right)$$

$$m_{x\theta}^{12} = -\left(2 - \frac{n^2+\frac{3}{2}+\nu}{a^2\beta^2}\right)$$

$$m_{x\theta}^{21} = -\frac{n\sqrt{n^2-1}}{a^2\beta^2}\frac{n^2-1}{n^2}\left(1 + \frac{1}{2}\frac{n^2-\frac{1}{2}-\nu}{a^2\beta^2}\right)$$

$$m_{x\theta}^{22} = -\frac{n\sqrt{n^2-1}}{a^2\beta^2}\frac{n^2-1}{n^2}\left(1 - \frac{1}{2}\frac{n^2-\frac{1}{2}-\nu}{a^2\beta^2}\right)$$

$$v_x^{11} = -\left(2 - \frac{2n^2-\frac{3}{2}+\nu}{a^2\beta^2}\right)$$

$$v_x^{12} = \left(2 + \frac{2n^2-\frac{3}{2}+\nu}{a^2\beta^2}\right)$$

$$v_x^{21} = \frac{n^2-1}{a^2\beta^2}\frac{n\sqrt{n^2-1}}{2a^2\beta^2}$$

$$v_x^{22} = \frac{n^2-1}{a^2\beta^2}\frac{n\sqrt{n^2-1}}{2a^2\beta^2}$$

$$v_x^{*11} = -\left(2 - \frac{n^2-\frac{3}{2}+\nu(n^2+1)}{a^2\beta^2}\right)$$

$$v_x^{*12} = \left(2 + \frac{n^2-\frac{3}{2}+\nu(n^2+1)}{a^2\beta^2}\right)$$

$$v_x^{*21} = -\frac{2-\nu}{2}\frac{n^2-1}{a^2\beta^2} - \frac{n\sqrt{n^2-1}}{a^2\beta^2}$$

$$v_x^{*22} = -\frac{2-\nu}{2}\frac{n^2-1}{a^2\beta^2} - \frac{n\sqrt{n^2-1}}{a^2\beta^2}$$

$$n_{x\theta}^{11} = -2\left(1 - \frac{1}{2}\frac{n^2+\frac{1}{2}}{a^2\beta^2}\right)$$

$$n_{x\theta}^{12} = 2\left(1 + \frac{1}{2}\frac{n^2+\frac{1}{2}}{a^2\beta^2}\right)$$

$$n_{x\theta}^{21} = -\frac{n^2-1}{n^2}\frac{n\sqrt{n^2-1}}{a^2\beta^2}\left(1 - \frac{1}{2}\frac{n^2+\frac{1}{2}}{a^2\beta^2}\right)$$

$$n_{x\theta}^{22} = -\frac{n^2-1}{n^2}\frac{n\sqrt{n^2-1}}{a^2\beta^2}\left(1 + \frac{1}{2}\frac{n^2+\frac{1}{2}}{a^2\beta^2}\right)$$

$$\eta = \frac{n\sqrt{n^2-1}}{a^2\beta^2}$$

$$\gamma = \frac{n^2-\frac{1}{2}}{a^2\beta^2}$$

Table 9.3 The approximated expressions (derived with the Donnell equation (9.1)) for the quantities of the homogeneous solution for the circular edge (note: $\gamma = n^2/(a\beta)^2$)

	Multiplicator	$e^{-[(1+\frac{1}{2}\gamma)\beta x]}\cos[(1-\frac{1}{2}\gamma)\beta x]$	$e^{-[(1+\frac{1}{2}\gamma)\beta x]}\sin[(1-\frac{1}{2}\gamma)\beta x]$	$e^{[(1+\frac{1}{2}\gamma)\beta x]}\cos[(1-\frac{1}{2}\gamma)\beta x]$	$e^{[(1+\frac{1}{2}\gamma)\beta x]}\sin[(1-\frac{1}{2}\gamma)\beta x]$
u_z	$\cos(n\theta)$	C_1	C_2	C_3	C_4
φ_x	$\dfrac{\beta}{2}\cos(n\theta)$	$(2+\gamma)C_1 - (2-\gamma)C_2$	$(2-\gamma)C_1 + (2+\gamma)C_2$	$-(2+\gamma)C_3 - (2-\gamma)C_4$	$(2-\gamma)C_3 - (2+\gamma)C_4$
u_x	$\dfrac{1}{4a\beta}\cos(n\theta)$	$(2\nu - \gamma(1+\nu))C_1$ $+(2\nu+\gamma(1+\nu))C_2$	$-(2\nu+\gamma(1+\nu))C_1$ $+(2\nu-\gamma(1+\nu))C_2$	$-(2\nu-\gamma(1+\nu))C_3$ $+(2\nu+\gamma(1+\nu))C_4$	$-(2\nu+\gamma(1+\nu))C_3$ $-(2\nu-\gamma(1+\nu))C_4$
u_θ	$\dfrac{1}{n}\sin(n\theta)$	$\gamma(1+\tfrac{1}{2}\nu)C_2$	$\gamma(1+\tfrac{1}{2}\nu)C_1$	$-\gamma(1+\tfrac{1}{2}\nu)C_4$	$\gamma(1+\tfrac{1}{2}\nu)C_3$
m_{xx}	$2D_b\beta^2\cos(n\theta)$	$-\gamma(1-\tfrac{1}{2}\nu)C_1 + C_2$	$-C_1 - \gamma(1-\tfrac{1}{2}\nu)C_2$	$-\gamma(1-\tfrac{1}{2}\nu)C_3 - C_4$	$C_3 - \gamma(1-\tfrac{1}{2}\nu)C_4$
$m_{\theta\theta}$	$D_b\beta^2\cos(n\theta)$	$\gamma(1-2\nu)C_1 + 2\nu C_2$	$-2\nu C_1 + \gamma(1-2\nu)C_2$	$\gamma(1-2\nu)C_3 - 2\nu C_4$	$2\nu C_3 + \gamma(1-2\nu)C_4$
$m_{x\theta}$	$D_b\beta\dfrac{1-\nu}{2}\dfrac{n}{a}\sin(n\theta)$	$-(2+\gamma)C_1 + (2-\gamma)C_2$	$-(2-\gamma)C_1 - (2+\gamma)C_2$	$(2+\gamma)C_3 + (2-\gamma)C_4$	$-(2-\gamma)C_3 + (2+\gamma)C_4$
v_x	$2D_b\beta^3\cos(n\theta)$	$-(1-\gamma)C_1 - (1+\gamma)C_2$	$(1+\gamma)C_1 - (1-\gamma)C_2$	$(1-\gamma)C_3 - (1+\gamma)C_4$	$(1+\gamma)C_3 + (1-\gamma)C_4$
v_θ	$D_b\beta^2\dfrac{n}{a}\sin(n\theta)$	$\gamma C_1 - 2C_2$	$2C_1 + \gamma C_2$	$\gamma C_3 + 2C_4$	$-2C_3 + \gamma C_4$
n_{xx}	$\dfrac{Et}{2a^2\beta^2}\left(\dfrac{n}{a}\right)^2 a\cos(n\theta)$	$-C_2$	C_1	C_4	$-C_3$
$n_{\theta\theta}$	$\dfrac{Et}{a}\cos(n\theta)$	$C_1 + \gamma C_2$	$-\gamma C_1 + C_2$	$C_3 - \gamma C_4$	$\gamma C_3 + C_4$
$n_{x\theta}$	$\dfrac{Et}{4a\beta}\dfrac{n}{a}\sin(n\theta)$	$-(2-\gamma)C_1 - (2+\gamma)C_2$	$(2+\gamma)C_1 - (2-\gamma)C_2$	$(2-\gamma)C_3 - (2+\gamma)C_4$	$(2+\gamma)C_3 + (2-\gamma)C_4$

Table 9.3 (continued)

	Multiplicator	$e^{-(\gamma\frac{\beta}{2}x)}\cos(\gamma\frac{\beta}{2}x)$	$e^{-(\gamma\frac{\beta}{2}x)}\sin(\gamma\frac{\beta}{2}x)$	$e^{(\gamma\frac{\beta}{2}x)}\cos(\gamma\frac{\beta}{2}x)$	$e^{(\gamma\frac{\beta}{2}x)}\sin(\gamma\frac{\beta}{2}x)$
u_z	$\cos(n\theta)$	C_5	C_6	C_7	C_8
φ_x	$\frac{\beta}{2}\cos(n\theta)$	$\gamma C_5 - \gamma C_6$	$\gamma C_5 + \gamma C_6$	$-\gamma C_7 - \gamma C_8$	$\gamma C_7 - \gamma C_8$
u_x	$\frac{1}{4a\beta}\cos(n\theta)$	$(2-\gamma(1+v))C_5$ $-(2+\gamma(1+v))C_6$	$(2+\gamma(1+v))C_5$ $+(2-\gamma(1+v))C_6$	$-(2-\gamma(1+v))C_7$ $-(2+\gamma(1+v))C_8$	$(2+\gamma(1+v))C_7$ $-(2-\gamma(1+v))C_8$
u_θ	$\frac{1}{n}\sin(n\theta)$	$-C_5 - \frac{1}{2}v\gamma C_6$	$\frac{1}{2}v\gamma C_5 - C_6$	$-C_7 + \frac{1}{2}v\gamma C_8$	$-\frac{1}{2}v\gamma C_7 - C_8$
m_{xx}	$2D_b\beta^2\cos(n\theta)$	$\frac{1}{2}v\gamma C_5$	$\frac{1}{2}v\gamma C_6$	$\frac{1}{2}v\gamma C_7$	$\frac{1}{2}v\gamma C_8$
$m_{\theta\theta}$	$D_b\beta^2\cos(n\theta)$	γC_5	γC_6	γC_7	γC_8
$m_{x\theta}$	$-D_b\beta\frac{1-vn}{2}\frac{n}{a}\sin(n\theta)$	$-\gamma C_5 + \gamma C_6$	$-\gamma C_5 - \gamma C_6$	$\gamma C_7 + \gamma C_8$	$-\gamma C_7 + \gamma C_8$
v_x	$2D_b\beta^3\cos(n\theta)$	0	0	0	0
v_θ	$D_b\beta^2\frac{n}{a}\sin(n\theta)$	$-\gamma C_5$	$-\gamma C_6$	$-\gamma C_7$	$-\gamma C_8$
n_{xx}	$\frac{Et}{2a^2\beta^2}\left(\frac{n}{a}\right)^2 a\cos(n\theta)$	C_6	$-C_5$	$-C_6$	C_5
$n_{\theta\theta}$	$\frac{Et}{a}\cos(n\theta)$	0	0	0	0
$n_{x\theta}$	$\frac{Et}{4a\beta}\frac{n}{a}\sin(n\theta)$	$\gamma C_5 + \gamma BC_6$	$-\gamma C_5 + \gamma C_6$	$-\gamma C_7 + \gamma C_8$	$-\gamma C_7 - \gamma C_8$

By substitution of the expressions for the displacement u_z into the relevant expressions, we obtain the inhomogeneous solution for all nontrivial quantities.

9.8 Complete Solution for Self-Balancing Modes

Describing a linear load p_z by the form

$$p_{zn}(x) = p_{zn}^{(2)}\frac{x}{l} + p_{zn}^{(1)}, \tag{9.65}$$

the complete solution for displacement u_z reads

$$
\begin{aligned}
u_z(x,\theta) = \cos n\theta \Big\{ &e^{-a_n^1\beta x}\big[C_1^n\cos(b_n^1\beta x) + C_2^n\sin(b_n^1\beta x)\big] \\
&+ e^{a_n^1\beta x}\big[C_3^n\cos(b_n^1\beta x) + C_4^n\sin(b_n^1\beta x)\big] \\
&+ e^{-a_n^2\beta x}\big[C_5^n\cos(b_n^2\beta x) + C_6^n\sin(b_n^2\beta x)\big] \\
&+ e^{a_n^2\beta\frac{x}{a}}\big[C_7^n\cos(b_n^2\beta x) + C_8^n\sin(b_n^2\beta x)\big] \Big\} \\
&+ \frac{1}{D_b}\left(\frac{a^2}{n^2-1}\right)^2\cos\theta\left(p_{zn}^{(2)}\frac{x}{l} + p_{zn}^{(1)}\right)
\end{aligned}
\tag{9.66}
$$

We obtain similar expressions for the independent displacements u_θ and u_x, and all other quantities of interest, by the appropriate substitutions.

9.9 Complete Solution for the Axisymmetric Load and Beam Load

9.9.1 Axisymmetric Mode

For $n = 0$, Eq. (9.34) reduces to

$$\left[\left(\frac{d^2}{dx^2} + \frac{1}{a^2}\right)^2 + 4\beta^4\right]u_{zn}(x) = \frac{1}{D_b}p_{zn}(x). \tag{9.67}$$

The complete solution is introduced without elaboration and reads

$$
\begin{aligned}
u_z(x) = &e^{-a_0\beta x}[C_1\cos(b_0\beta x) + C_2\sin(b_0\beta x)] \\
&+ e^{a_0\beta x}[C_3\cos(b_0\beta x) + C_4\sin(b_0\beta x)] \\
&+ \frac{1}{Et}\left[a^2\left(p_{z0}^{(2)}\frac{x}{l} + p_{z0}^{(1)}\right) + va(p_x x + C_5)\right]
\end{aligned}
\tag{9.68}
$$

in which the loads p_x and p_z are described by the forms

$$p_{x0}(x) = p_{x0}$$
$$p_{z0}(x) = p_{z0}^{(2)} \frac{x}{l} + p_{z0}^{(1)} \tag{9.69}$$

and l represents the length of the circular cylinder.

The inhomogeneous solution is identical to Eq. (4.57) for the membrane behaviour and, as shown by roots of the homogeneous solution in Eq. (9.41) and the influence length of Eq. (9.45), the homogeneous solution is equivalent to the bending behaviour as obtained in Chap. 5.

9.9.2   Beam Mode

For $n = 1$, Eq. (9.34) reduces to

$$\left[\left(\frac{d^2}{dx^2} - \frac{1}{a^2} \right)^2 + 4\beta^4 \right] \frac{d^4}{dx^4} u_{zn}(x) = \frac{1}{D_b} \left(\frac{d^2}{dx^2} - \frac{1}{a^2} \right)^2 p_{zn}(x). \tag{9.70}$$

The complete solution is introduced without elaboration and read

$$u_z(x,\theta) = \begin{matrix} \cos\theta \{ e^{-a_1\beta x} [C_1 \cos(b_1) + C_2 \sin(b_1\beta x)] \\ + e^{a_1\beta x} [C_3 \cos(b_1\beta x) + C_4 \sin(b_1)] \} \end{matrix}$$
$$+ \frac{1}{Eta^2} \cos\theta \left\{ \frac{p_{z1}^{(2)}}{l} \left(\frac{1}{120} x^5 - \frac{1}{3} a^2 x^3 + a^4 x \right) + p_{z1}^{(1)} \left(\frac{1}{24} x^4 - a^2 x^2 + a^4 \right) \right.$$
$$- \frac{p_{\theta 1}^{(2)}}{l} \left(\frac{1}{120} x^5 - \frac{2+v}{6} a^2 x^3 \right) - p_{\theta 1}^{(1)} \left(\frac{1}{24} x^4 - \frac{2+v}{2} a^2 x^2 \right)$$
$$\left. + a p_{x1} \left(\frac{1}{6} x^3 + v a^2 x \right) \right\}$$
$$+ \cos\theta \left\{ \left(\frac{1}{6} \left(\frac{x}{a} \right)^3 - 2\frac{x}{a} \right) C_5 + \left(\frac{1}{2} \left(\frac{x}{a} \right)^2 - 2 \right) C_6 + \frac{x}{a} C_7 + C_8 \right\} \tag{9.71}$$

in which the loads p_x, p_θ and p_z are described by the forms

$$p_{x1}(x) = p_{x1}$$
$$p_{\theta 1}(x) = p_{\theta 1}^{(2)} \frac{x}{l} + p_{\theta 1}^{(1)}$$
$$p_{z1}(x) = p_{z1}^{(2)} \frac{x}{l} + p_{z1}^{(1)}. \tag{9.72}$$

and l represents the length of the circular cylinder. The inhomogeneous solution is identical to Eq. (4.17) for the membrane behaviour. As shown by roots of the homogeneous solution in Eq. (9.41) and the influence length of Eq. (9.45), the homogeneous solution describes a bending behaviour that describes a short edge disturbance. For long circular cylindrical shells under beam load, we have described in Sect. 4.6 that the incompatibility of the membrane solution at the edges need to be compensated by bending acting. We conclude that these bending moments only exert influence within a short distance from the edge and that these moments have a negligible effect on the total stress within this area. This observation is strengthened by the calculations as detailed in Chap. 12, where long chimneys are treated, in which the range of application of the above is assessed. In Chap. 13, where storage tanks are treated, it is shown that for such short circular cylindrical shells under beam load, the edge disturbances cannot be neglected.

References

1. Love AEH (1927) The mathematical theory of elasticity, 4th edn. Cambridge University Press, Cambridge
2. Flügge W (1960) Stresses in shells. Springer, Berlin (corrected reprint 1962)
3. Wlassow WS (1958) Algemeine Schalentheorie und ihere Anwendung in der Technik. Akademie Verlag, Berlin
4. Novoshilov VV (1959) The theory of thin shells. Translated from the Russian (1951), ed by PG Lowe, Groningen, Noordhoff
5. Reissner E (1966) On the derivation of the theory of elastic shells. In: Proceedings 11th international congress of applied mechanics, Munich 1964. Springer, Berlin, pp 20–20
6. Morley LSD (1959) An improvement on Donnell's approximation for thin-walled circular cylinders. Q J Mech Appl Mech 12:88–99
7. Hoefakker JH (2010) Theory review for cylindrical shells and parametric study of chimneys and tanks. Doctoral thesis, Delft University of Technology. Eburon Academic Publishers, Delft
8. Niordson FI (1985) Shell theory. North-Holland series in applied mechanics and mechanics. Elsevier Science, New York

Chapter 10
Semi-Membrane Concept Theory for Circular Cylindrical Shells

In the previous chapter, we learned about short and long influence lengths for full circular cylindrical shells. For axisymmetric loading and beam-type loading, only the short influence length plays a role, and the edge disturbance is governed by a fourth-order differential equation. For the self-balancing modes, both the short and long influence lengths play a role, and the more complex eighth-order differential equation of the Morley theory is needed. It is possible for this loading type to reduce also to a fourth-order differential equation, an approximation in which only the long influence length will play a role. This so-called *semi-membrane concept (SMC)*, introduced by Pircher et al. [1], is the subject of the present chapter.

10.1 Introduction

For convenience and without degenerating the generality of the approach, we restrict the applications in the present chapter to the same cases as treated in Chap. 9. We consider full circular cylindrical shells subject only to loading normal to its surface, which is symmetric with respect to the axis $\theta = 0$, constant or linear in axial direction, and with boundary conditions that are specified at the circular edges.

The semi-membrane concept assumes that, to simplify the initial kinematical equations, the circumferential strain $\varepsilon_{\theta\theta}$ is equal to zero and that, to simplify the initial equilibrium equations, the axial bending moment m_{xx} and the twisting moment $m_{x\theta}$ are zero, and hence the transverse shear force v_x. The zero strain $\varepsilon_{\theta\theta}$ is accompanied by a negligible membrane force $n_{\theta\theta}$ and the zero moments m_{xx} and $m_{x\theta}$ by negligible curvatures κ_{xx} and $\rho_{x\theta}$. The resulting equation exactly describes the ring-bending behaviour of self-balancing modes. We can apply this concept to, e.g., a radial wind load, an axial elastic support at the circular edge and an axial support displacement. As we did in the previous chapter, we assume that the parameters describing the material properties and the cross-sectional geometry, i.e., E, ν and a, t respectively, are constant for the whole circular cylindrical shell.

J. Blaauwendraad and J. H. Hoefakker, *Structural Shell Analysis*,
Solid Mechanics and Its Applications 200, DOI: 10.1007/978-94-007-6701-0_10,
© Springer Science+Business Media Dordrecht 2014

10.2 Sets of Equations

For the semi-membrane concept, we apply the same polar coordinate system and choose the axes in the same direction as adopted for the Morley theory in Sect. 9.1. It holds again that $k_x = 0$, $k_y = -1/a$ and $k_{xy} = 0$. The sets of equations formulated for the circular cylindrical shell are tremendously simplified by the above-mentioned assumptions of the semi-membrane concept. It should be noted that $a\varepsilon_{\theta\theta} = \partial u_\theta/\partial\theta + u_z = 0$ and hence

$$u_z = -\frac{\partial u_\theta}{\partial \theta} \tag{10.1}$$

This means that either u_z or u_θ is no longer an independent variable. Here we choose the displacement u_θ as independent. Hence, we must derive differential equations for the two displacements u_x and u_θ. Therefore, we have to transform the load p_z to a load p_θ, which is statically equivalent. If we substitute $-\partial u_\theta/\partial\theta$ for displacement u_z, we must replace load p_z by a load in u_θ-direction of the size

$$p_\theta = \partial p_z/\partial\theta \tag{10.2}$$

which can be shown on the basis of virtual work considerations. As a result, the scheme of relations of Fig. 10.1 applies.

We obtain the kinematical equation by rewriting Eq. (9.14) in two steps. First we skip the rows that correspond with the zero strain $\varepsilon_{\theta\theta}$ and negligible curvatures κ_{xx} and $\rho_{x\theta}$:

$$\begin{bmatrix} \varepsilon_{xx} \\ \gamma_{x\theta} \\ \kappa_{\theta\theta} \end{bmatrix} = \begin{bmatrix} \dfrac{\partial}{\partial x} & 0 & 0 \\ \dfrac{1}{a}\dfrac{\partial}{\partial\theta} & \dfrac{\partial}{\partial x} & 0 \\ 0 & 0 & -\dfrac{1}{a^2}\dfrac{\partial^2}{\partial\theta^2} - \dfrac{1}{a^2} \end{bmatrix} \begin{bmatrix} u_x \\ u_\theta \\ u_z \end{bmatrix} \tag{10.3}$$

Next, we make use of the introduced relation $u_z = -\partial u_\theta/\partial\theta$, and rewrite Eq. (10.3) to

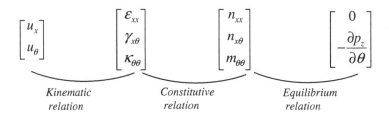

$$\begin{bmatrix} u_x \\ u_\theta \end{bmatrix} \qquad \begin{bmatrix} \varepsilon_{xx} \\ \gamma_{x\theta} \\ \kappa_{\theta\theta} \end{bmatrix} \qquad \begin{bmatrix} n_{xx} \\ n_{x\theta} \\ m_{\theta\theta} \end{bmatrix} \qquad \begin{bmatrix} 0 \\ -\dfrac{\partial p_z}{\partial\theta} \end{bmatrix}$$

$$\underbrace{\hphantom{XXXXX}}_{\substack{\textit{Kinematic} \\ \textit{relation}}} \qquad \underbrace{\hphantom{XXXXX}}_{\substack{\textit{Constitutive} \\ \textit{relation}}} \qquad \underbrace{\hphantom{XXXXX}}_{\substack{\textit{Equilibrium} \\ \textit{relation}}}$$

Fig. 10.1 Scheme of relationships in SMC-theory

$$
\begin{bmatrix} \varepsilon_{xx} \\ \gamma_{x\theta} \\ \kappa_{\theta\theta} \end{bmatrix} = \begin{bmatrix} \dfrac{\partial}{\partial x} & 0 \\ \dfrac{1}{a}\dfrac{\partial}{\partial\theta} & \dfrac{\partial}{\partial x} \\ 0 & \dfrac{1}{a^2}\dfrac{\partial^3}{\partial\theta^3} + \dfrac{1}{a^2}\dfrac{\partial}{\partial\theta} \end{bmatrix} \begin{bmatrix} u_x \\ u_\theta \end{bmatrix} \tag{10.4}
$$

in which only the two independent displacements are employed.

The constitutive relation (9.15), rewritten for the assumptions introduced above, transforms into

$$
\begin{bmatrix} n_{xx} \\ n_{x\theta} \\ m_{\theta\theta} \end{bmatrix} = \begin{bmatrix} D_m(1-v^2) & 0 & 0 \\ 0 & D_s & 0 \\ 0 & 0 & D_b \end{bmatrix} \begin{bmatrix} \varepsilon_{xx} \\ \gamma_{x\theta} \\ \kappa_{\theta\theta} \end{bmatrix} \tag{10.5}
$$

where the quantities D_m, D_s and D_b are the membrane rigidity, the shear rigidity and the flexural rigidity, respectively, which are given by

$$
D_m = \frac{Et}{1-v^2}; \quad D_s = \frac{Et}{2(1+v)}; \quad D_b = \frac{Et^3}{12(1-v^2)} \tag{10.6}
$$

In order to obtain the equilibrium relation (9.18), we made use of the fact that the differential operator is the adjoint of the differential operator in the kinematic relation. Applying this again to Eq. (10.4), the equilibrium equation now becomes

$$
\begin{bmatrix} -\dfrac{\partial}{\partial x} & -\dfrac{1}{a}\dfrac{\partial}{\partial\theta} & 0 \\ 0 & -\dfrac{\partial}{\partial x} & -\dfrac{1}{a^2}\dfrac{\partial^3}{\partial\theta^3} - \dfrac{1}{a^2}\dfrac{\partial}{\partial\theta} \end{bmatrix} \begin{bmatrix} n_{xx}a \\ n_{x\theta}a \\ m_{\theta\theta}a \end{bmatrix} = \begin{bmatrix} 0 \\ a\dfrac{\partial p_z}{\partial\theta} \end{bmatrix} \tag{10.7}
$$

The transverse shear force, described by Eq. (9.19), becomes

$$
v_\theta = \frac{1}{a}\frac{\partial m_{\theta\theta}}{\partial\theta} \tag{10.8}
$$

The boundary conditions (9.23) are rewritten on the basis of application of the principle of virtual work to the kinematical relation (10.4). Without providing further elaboration, we obtain for curved edges

$$
\begin{array}{l} - \text{ either } u_x \text{ or } n_{xx} \\ - \text{ either } u_\theta \text{ or } n_{x\theta} \end{array} \tag{10.9}
$$

10.3 Differential Equations for Load p_z

10.3.1 The Differential Equations for Displacements

Substitution of the kinematical relation (10.4) into the constitutive relation (10.5) results in what is sometimes referred to as the "elastic law":

$$n_{xx} = D_m(1 - v^2)\frac{\partial u_x}{\partial x}$$

$$n_{x\theta} = D_s\left(\frac{1}{a}\frac{\partial u_x}{\partial \theta} + \frac{\partial u_\theta}{\partial x}\right) \tag{10.10}$$

$$m_{\theta\theta} = D_b\left(\frac{1}{a^2}\frac{\partial^3 u_\theta}{\partial \theta^3} + \frac{1}{a^2}\frac{\partial u_\theta}{\partial \theta}\right)$$

Substitution of this elastic law into Eq. (10.7) yields the following two differential equations for the displacements

$$-(1 - v^2)\frac{\partial^2 u_x}{\partial x^2} - \frac{D_s}{D_m}\left(\frac{1}{a^2}\frac{\partial^2 u_x}{\partial \theta^2} + \frac{1}{a}\frac{\partial^2 u_\theta}{\partial x\partial\theta}\right) = 0$$

$$\tag{10.11}$$

$$-\frac{D_s}{D_m}\left(\frac{1}{a}\frac{\partial^2 u_x}{\partial x\partial\theta} + \frac{\partial^2 u_\theta}{\partial x^2}\right) - \frac{D_b}{D_m}\frac{1}{a^4}\frac{\partial^2}{\partial\theta^2}\left(\frac{\partial^2}{\partial\theta^2} + 1\right)^2 u_\theta = \frac{1}{D_m}\frac{\partial p_z}{\partial\theta}$$

Symbolically, we describe them by

$$\begin{bmatrix} L_{11} & L_{12} \\ L_{21} & L_{22} \end{bmatrix}\begin{bmatrix} u_x \\ u_\theta \end{bmatrix} = -\frac{1}{D_m}\begin{bmatrix} 0 \\ \frac{\partial p_z}{\partial\theta} \end{bmatrix} \tag{10.12}$$

The operators L_{11} up to and including L_{22} form a differential operator matrix, in which the operators are

$$L_{11} = (1 - v^2)\frac{\partial^2}{\partial x^2} + \frac{1 - v}{2}\frac{1}{a^2}\frac{\partial^2}{\partial\theta^2}$$

$$L_{12} = L_{21} = \frac{1 - v}{2}\frac{1}{a}\frac{\partial^2}{\partial x\partial\theta} \tag{10.13}$$

$$L_{22} = \frac{1 - v}{2}\frac{\partial^2}{\partial x^2} + \frac{k}{a^2}\frac{\partial^2}{\partial\theta^2}\left(\frac{\partial^2}{\partial\theta^2} + 1\right)^2$$

Here the dimensionless parameter k is identical to the parameter defined by Eq. (9.28):

$$k = \frac{D_b}{D_m a^2} = \frac{t^2}{12a^2} \tag{10.14}$$

For a thin shell where $t < a$, the parameter k is negligibly small in comparison to unity ($k \ll 1$).

10.3.2 The Single Differential Equation

By eliminating u_x from the two equations, we obtain the single differential equation for the displacement u_θ, which symbolically reads

$$(L_{11}L_{22} - L_{21}L_{12})u_\theta = -\frac{L_{11}}{D_m}\frac{\partial p_z}{\partial \theta} \tag{10.15}$$

We then obtain the single differential equation

$$\left[\frac{\partial^4}{\partial x^4} + \frac{k}{a^2(1-v^2)}\left(2(1+v)\frac{\partial^2}{\partial x^2} + \frac{1}{a^2}\frac{\partial^2}{\partial \theta^2}\right)\frac{\partial^2}{\partial \theta^2}\left(\frac{\partial^2}{\partial \theta^2} + 1\right)^2\right]u_\theta$$

$$= -\frac{1}{D_m(1-v^2)}\left(2(1+v)\frac{\partial^2}{\partial x^2} + \frac{1}{a^2}\frac{\partial^2}{\partial \theta^2}\right)\frac{\partial p_z}{\partial \theta} \tag{10.16}$$

To facilitate comparison between the solutions presented herein, we prefer to solve the homogeneous equation for the displacement u_z, as then all quantities for the semi-membrane concept can be described similar to those resulting from the solution to the Morley equation. This can be easily accomplished by noting that the relation $u_z = -\partial u_\theta/\partial \theta$ holds. Hence, by taking the derivative of Eq. (10.16) with respect to θ and by rearranging the resulting *SMC-equation*, the single differential equation for u_z becomes

$$\left[4\beta^4\frac{\partial^4}{\partial x^4} + \frac{1}{a^8}\left(2(1+v)a^2\frac{\partial^2}{\partial x^2} + \frac{\partial^2}{\partial \theta^2}\right)\frac{\partial^2}{\partial \theta^2}\left(\frac{\partial^2}{\partial \theta^2} + 1\right)^2\right]u_z$$

$$= \frac{1}{D_b}\left(2(1+v)\frac{1}{a^2}\frac{\partial^4}{\partial x^2\partial \theta^2} + \frac{1}{a^4}\frac{\partial^4}{\partial \theta^4}\right)p_z \tag{10.17}$$

Here the parameter β is identical to the parameter defined by Eq. (9.3):

$$\beta^4 = \frac{3(1-v^2)}{(at)^2} \tag{10.18}$$

In comparison with Morley's equation in Eq. (9.31), the change seems dramatic. Hereafter, in the present chapter, we will see that the impact of this tremendous reduction is minor on the overall response of full circular cylindrical shells subject to loads associated with the self-balancing modes.

10.4 Homogeneous Solution of the Differential Equation for a Curved Edge

As stated in Sect. 10.1, the semi-membrane concept exactly describes the ring-bending behavior and can only be applied to self-balancing modes. The modes indicated by $n = 2, 3, 4, \ldots$ are generally known as these self-balancing modes.

For these mode numbers $n > 1$, all quantities can be expressed as functions of the type as adopted in Sect. 9.4.1. Hence, we can make the following substitutions for the loads and displacements associated with self-balancing modes

$$
\begin{aligned}
p_x(x, \theta) &= 0; & u_x(x, \theta) &= u_{xn}(x) \cos n\theta \\
p_\theta(x, \theta) &= 0; & u_\theta(x, \theta) &= u_{\theta n}(x) \sin n\theta \\
p_z(x, \theta) &= p_{zn}(x) \cos n\theta; & u_z(x, \theta) &= u_{zn}(x) \cos n\theta
\end{aligned}
\tag{10.19}
$$

while for the derivatives with respect to the circumferential coordinate θ we can make substitutions of the form

$$
\frac{\partial \phi(x, \theta)}{\partial \theta} = \phi_n(x) \frac{\partial \cos n\theta}{\partial \theta} = -n\phi_n(x) \sin n\theta
\tag{10.20}
$$

for quantities generally described by $\phi(x, \theta) = \phi_n(x) \cos n\theta$ and similarly for the quantities generally described by $\phi(x, \theta) = \phi_n(x) \sin n\theta$.

By substitution of the load and displacement functions given above, the SMC-equation (10.17) becomes an ordinary differential equation and by omitting the cosine function for the circumferential distribution, the governing differential equation is reduced to

$$
\begin{aligned}
&\left[4\beta^4 \frac{d^4}{dx^4} - \frac{n^2}{a^2} \left(2(1 + v) \frac{d^2}{dx^2} - \frac{n^2}{a^2} \right) \left(\frac{n^2 - 1}{a^2} \right)^2 \right] u_{zn}(x) \\
&\qquad = \frac{1}{D_b} \left(-2(1 + v) \frac{n^2}{a^2} \frac{\partial^2}{\partial x^2} + \frac{n^4}{a^4} \right) p_{zn}(x)
\end{aligned}
\tag{10.21}
$$

The homogeneous equation is given by

$$
\left[\frac{d^4}{dx^4} - \frac{1 + v}{2} \frac{n^2 (n^2 - 1)^2}{a^2 (a\beta)^4} \frac{d^2}{dx^2} + \frac{n^4 (n^2 - 1)^2}{a^4 \, 4(a\beta)^4} \right] u_{zn}(x) = 0
\tag{10.22}
$$

Identical to the substitution in Sect. 9.4.1, the periodic trial function for $u_z(x, \theta)$ is $u_z(x, \theta) = F_n(x) \cos n\theta$. We obtain the solution:

$$
\begin{aligned}
u_{zn}(x) &= e^{-a_n^{SMC} \beta x} \left[C_1^n \cos\left(b_n^{SMC} \beta x \right) + C_2^n \sin\left(b_n^{SMC} \beta x \right) \right] \\
&\quad + e^{a_n^{SMC} \beta x} \left[C_3^n \cos\left(b_n^{SMC} \beta x \right) + C_4^n \sin\left(b_n^{SMC} \beta x \right) \right]
\end{aligned}
\tag{10.23}
$$

where the dimensionless roots a_n^{SMC} and b_n^{SMC} are defined by

$$
a_n^{SMC} = \frac{1}{2} \eta_n \sqrt{1 + \gamma_n^{SMC}}, \qquad b_n^{SMC} = \frac{1}{2} \eta_n \sqrt{1 - \gamma_n^{SMC}}
\tag{10.24}
$$

in which

$$
\gamma_n^{SMC} = \frac{1}{2}(1 + v)(n^2 - 1) \big/ (a\beta)^2, \qquad \eta_n = n(n^2 - 1)^{\frac{1}{2}} \big/ (a\beta)^2
\tag{10.25}
$$

Note that only four integration constants C_i occur, where in the solution of Morley's equation in Eq. (9.36) eight constants appear.

Similar to the approximation in Sect. 9.4.2, we obtain the approximate expressions

$$a_n^{SMC} = \frac{1}{2}\eta_n\left(1 + \frac{1}{2}\gamma_n^{SMC}\right), \quad b_n^{SMC} = \frac{1}{2}\eta_n\left(1 - \frac{1}{2}\gamma_n^{SMC}\right) \tag{10.26}$$

which are in full correspondence with the roots as defined by Eq. (9.42) that have been associated with the long-wave solution in Sect. 9.5.3.

10.5 Influence Length

Identical to the determination of the influence length in Sect. 9.5.3, the characteristic and influence lengths for the long-wave solution are approximated by

$$l_{c,2} \approx \frac{2}{\eta_n\beta} = \frac{2\sqrt[4]{3(1-v^2)}}{n\sqrt{n^2-1}}\frac{a}{t}\sqrt{at}; \quad l_{i,2} \approx \frac{8.1}{n\sqrt{n^2-1}}\frac{a}{t}\sqrt{at} \tag{10.27}$$

This long influence length describes a far-reaching influence, which decreases rapidly with increasing n.

In the next sections, we will observe that the solutions for the displacements and shell forces are almost identical to those of the long-wave solution obtained from Morley's equation.

10.6 Displacements and Shell Forces of the Homogeneous Solution for the Self-Balancing Modes

Now that we have a homogeneous solution for the displacement u_z, it is necessary to express the other displacements, membrane forces, moments and transverse shear forces as functions of u_z. We can do so for the displacement u_θ by solving the relation $u_z = -\partial u_\theta/\partial\theta$, which leads to

$$u_\theta = -\int u_z d\theta \tag{10.28}$$

We obtain the solution for the displacement u_x by solving the second equation of the set (10.11):

$$\frac{1}{a}\frac{\partial^2 u_x}{\partial x\partial\theta} + \frac{\partial^2 u_\theta}{\partial x^2} + \frac{1+v}{2(a\beta)^4}\frac{1}{a^2}\frac{\partial^2}{\partial\theta^2}\left(\frac{\partial^2}{\partial\theta^2}+1\right)^2 u_\theta = 0 \tag{10.29}$$

By working out integrations with respect to x and θ, we rewrite this equation to

$$u_x = -\int \frac{\partial u_\theta}{\partial x}\,ad\theta - \frac{1+\nu}{2}\frac{1}{a(a\beta)^4}\int \frac{\partial}{\partial\theta}\left(\frac{\partial^2}{\partial\theta^2}+1\right)^2 u_\theta dx \qquad (10.30)$$

Upon substitution of Eq. (10.28) for the displacement u_θ, this expression reads

$$u_x = \int\int \frac{\partial u_z}{\partial x}\,ad\theta d\theta + \frac{1+\nu}{2}\frac{1}{a(a\beta)^4}\int\left(\frac{\partial^2}{\partial\theta^2}+1\right)^2 u_z dx \qquad (10.31)$$

By substitution of the displacement functions given above, this becomes an ordinary differential equation in which we omit the sine function (for u_θ) and the cosine function (for u_x).

The membrane forces and the moment are described by Eq. (10.10):

$$n_{xx} = D_m\left(1 - \nu^2\right)\frac{\partial u_x}{\partial x}$$

$$n_{x\theta} = D_s\left(\frac{1}{a}\frac{\partial u_x}{\partial\theta} + \frac{\partial u_\theta}{\partial x}\right) \qquad (10.32)$$

$$m_{\theta\theta} = D_b\left(\frac{1}{a^2}\frac{\partial^3 u_\theta}{\partial\theta^3} + \frac{1}{a^2}\frac{\partial u_\theta}{\partial\theta}\right)$$

Upon substitution of the expressions for the displacements above, these expressions read

$$n_{xx} = D_m\left(1 - \nu^2\right)\left(\int\int \frac{\partial^2 u_z}{\partial x^2}\,ad\theta d\theta + \frac{1+\nu}{2}\frac{1}{a(a\beta)^4}\left(\frac{\partial^2}{\partial\theta^2}+1\right)^2 u_z\right)$$

$$n_{x\theta} = D_s\left(\frac{1+\nu}{2}\frac{1}{a^2(a\beta)^4}\int \frac{\partial}{\partial\theta}\left(\frac{\partial^2}{\partial\theta^2}+1\right)^2 u_z dx\right) \qquad (10.33)$$

$$m_{\theta\theta} = -D_b\frac{1}{a^2}\left(\frac{\partial^2 u_z}{\partial\theta^2} + u_z\right)$$

The transverse shear force is described by Eq. (10.8) and becomes

$$v_\theta = \frac{1}{a}\frac{\partial m_{\theta\theta}}{\partial\theta} \qquad (10.34)$$

Upon substitution of the expression for the moment above, this expression reads

$$v_\theta = -D_b\frac{1}{a^3}\left(\frac{\partial^3 u_z}{\partial\theta^3} + \frac{\partial u_z}{\partial\theta}\right) \qquad (10.35)$$

By working out the derivatives and integrals, we obtain all displacements and shell forces as functions of the coordinates x and θ multiplied by the four constants of the homogeneous solution. The resulting expressions are presented in Table 10.1.

Table 10.1 The expressions (based on Eq. (10.26)) for the quantities of the homogeneous solution for the circular edge

	Multiplicator	$e^{-\left[\eta\left(1+\frac{1}{2}\gamma\right)\frac{\beta}{2}x\right]}\cos\left[\eta\left(1-\frac{1}{2}\gamma\right)\frac{\beta}{2}x\right]$	$e^{-\left[\eta\left(1+\frac{1}{2}\gamma\right)\frac{\beta}{2}x\right]}\sin\left[\eta\left(1-\frac{1}{2}\gamma\right)\frac{\beta}{2}x\right]$	$e^{\left[\eta\left(1+\frac{1}{2}\gamma\right)\frac{\beta}{2}x\right]}\cos\left[\eta\left(1-\frac{1}{2}\gamma\right)\frac{\beta}{2}x\right]$	$e^{\left[\eta\left(1+\frac{1}{2}\gamma\right)\frac{\beta}{2}x\right]}\sin\left[\eta\left(1-\frac{1}{2}\gamma\right)\frac{\beta}{2}x\right]$
u_z	$\cos(n\theta)$	C_1	C_2	C_3	C_4
φ_x	$\dfrac{\beta}{2}\cos(n\theta)$	$\eta\left(1-\tfrac{1}{2}\gamma\right)C_1+\eta\left(1+\tfrac{1}{2}\gamma\right)C_2$	$\eta\left(1+\tfrac{1}{2}\gamma\right)C_1-\eta\left(1-\tfrac{1}{2}\gamma\right)C_2$	$-\eta\left(1+\tfrac{1}{2}\gamma\right)C_3-\left(1-\tfrac{1}{2}\gamma\right)C_4$	$\left(1-\tfrac{1}{2}\gamma\right)C_3-\left(1+\tfrac{1}{2}\gamma\right)C_4$
u_x	$\dfrac{1}{2\beta a}\dfrac{n\sqrt{n^2-1}}{n^2}\cos(n\theta)$	$\left(1-\tfrac{3}{2}\gamma\right)C_1-\left(1+\tfrac{3}{2}\gamma\right)C_2$	$\left(1+\tfrac{3}{2}\gamma\right)C_1+\left(1-\tfrac{3}{2}\gamma\right)C_2$	$-\left(1-\tfrac{3}{2}\gamma\right)C_3-\left(1+\tfrac{3}{2}\gamma\right)C_4$	$\left(1+\tfrac{3}{2}\gamma\right)C_3-\left(1-\tfrac{3}{2}\gamma\right)C_4$
u_θ	$\dfrac{1}{n}\sin(n\theta)$	$-C_1$	$-C_2$	$-C_3$	$-C_4$
$m_{\theta\theta}$	$D_b\dfrac{n^2-1}{a^2}\cos(n\theta)$	C_1	C_2	C_3	C_4
v_θ	$D_b\dfrac{n^2-1}{a^2}\dfrac{n}{a}\sin(n\theta)$	$-C_1$	$-C_2$	$-C_3$	$-C_4$
n_{xx}	$\dfrac{Et}{2(\beta a)^2}\dfrac{n^2-1}{n^2}\left(\dfrac{n}{a}\right)^2 a\cos(n\theta)$	γC_1+C_2	$-C_1+\gamma C_2$	γC_3-C_4	$C_3+\gamma C_4$
$n_{x\theta}$	$\dfrac{Et}{4\beta a}\dfrac{n^2-1}{n^2}\eta\dfrac{n}{a}\sin(n\theta)$	$\left(1+\tfrac{1}{2}\gamma\right)C_1+\left(1-\tfrac{1}{2}\gamma\right)C_2$	$-\left(1-\tfrac{1}{2}\gamma\right)C_1+\left(1+\tfrac{1}{2}\gamma\right)C_2$	$-\left(1+\tfrac{1}{2}\gamma\right)C_3+\left(1-\tfrac{1}{2}\gamma\right)C_4$	$-\left(1-\tfrac{1}{2}\gamma\right)C_3-\left(1+\tfrac{1}{2}\gamma\right)C_4$

10.7 Inhomogeneous Solution

Restricting ourselves to constant or linear loads p_z, we can reduce the solution to the inhomogeneous equation of Eq. (10.21) to

$$\frac{n^4}{a^4}\left(\frac{n^2-1}{a^2}\right)^2 u_{zn}(x) = \frac{1}{D_b}\frac{n^4}{a^4}p_{zn}(x) \qquad (10.36)$$

which is identical to the inhomogeneous equation of Eq. (9.61) for the Morley bending theory.

The inhomogeneous solution is

$$u_{zn}(x) = \frac{1}{D_b}\left(\frac{a^2}{n^2-1}\right)^2 p_{zn}(x) \qquad (10.37)$$

Similar to the homogeneous solution, the inhomogeneous solution for the displacement u_θ can be obtained by solving Eq. (10.28). The inhomogeneous solution for the displacement u_x can be obtained by solving the first expression of Eq. (10.11).

There is a good argument to omit the second derivative with respect to x in this expression. In the SMC-theory, the gradient in x-direction is an order of magnitude smaller than in θ-direction. Doing so, the latter differential equation becomes

$$\frac{1}{a^2}\frac{\partial^2 u_x}{\partial\theta^2} = -\frac{1}{a}\frac{\partial^2 u_\theta}{\partial x\partial\theta} \qquad (10.38)$$

We can rewrite this equation by substituting the displacement and load functions given above. If we omit the cosine and sine terms, the equations for u_θ and u_x become

$$\begin{aligned}
u_{\theta n}(x) &= -\frac{1}{n}u_{zn}(x) \\
u_{xn}(x) &= \frac{a}{n}\frac{du_{\theta n}(x)}{dx} = -\frac{a}{n^2}\frac{du_{zn}(x)}{dx}
\end{aligned} \qquad (10.39)$$

into which the solution of Eq. (10.37) can be substituted.

By substitution of the expressions for the displacements into the expressions of Eq. (10.10), we obtain the inhomogeneous solution for all nontrivial quantities.

10.8 Complete Solution

Describing a linear load p_z by the form

$$p_{zn}(x) = p_{zn}^{(2)}\frac{x}{l} + p_{zn}^{(1)} \qquad (10.40)$$

the complete solution for displacement u_z reads

$$u_z(x, \theta) = \theta \left\{ e^{-a_n^{SMC} \beta x} \left[C_1^n \cos\left(b_n^{SMC} \beta x\right) + C_2^n \sin\left(b_n^{SMC} \beta x\right) \right] \right.$$
$$\left. + e^{a_n^{SMC} \beta x} \left[C_3^n \cos\left(b_n^{SMC} \beta x\right) + C_4^n \sin\left(b_n^{SMC} \beta x\right) \right] \right\} \qquad (10.41)$$
$$+ \frac{1}{D_b} \left(\frac{a^2}{n^2 - 1}\right)^2 \cos n\theta \left(p_{zn}^{(2)} \frac{x}{l} + p_{zn}^{(1)}\right)$$

We obtain similar expressions for the independent displacements u_θ and u_x, and all other quantities of interest, by the appropriate substitutions.

10.9 Remark Considering Accuracy

As stated in Sect. 10.1, the SMC-equation can be applied to radial wind load, axial elastic supports and axial support displacements. Clearly, not all load cases or support conditions can be described. Moreover, the semi-membrane concept is only applicable to certain load-deformation behaviors of cylindrical shell structures. The simplifications presuppose that the influence of the part of the solution described by the short influence length can be neglected. This is the case, if the cylinder is sufficiently long compared with its radius, and if the boundary effects mainly influence the more distant material.

Consequently, the small terms of the dimensionless parameters a_n^{SMC} and b_n^{SMC} as presented by Eq. (10.26) are identified as superfluous and may be discarded so as to avoid suggesting an accuracy that is not described.

If we retain only the leading terms, by neglecting $(a\beta)^{-2}$ in comparison to unity, the dimensionless parameters a_n^{SMC} and b_n^{SMC} become equal to $\eta_n/2$. The complete solution is then described by

$$u_{zn}(x) = e^{-\frac{1}{2}\eta_n \beta x} \left[C_1^n \cos\left(\frac{1}{2}\eta_n \beta x\right) + C_2^n \sin\left(\frac{1}{2}\eta_n \beta x\right) \right]$$
$$+ e^{\frac{1}{2}\eta_n \beta x} \left[C_3^n \cos\left(\frac{1}{2}\eta_n \beta x\right) + C_4^n \sin\left(\frac{1}{2}\eta_n \beta x\right) \right] \qquad (10.42)$$
$$+ \frac{1}{D_b} \left(\frac{a^2}{n^2 - 1}\right)^2 p_{zn}(x)$$

and similarly the same approximation can be adopted for all other quantities.

Hence, it is readily verified that this approximated solution would be the exact solution to the following differential equation

$$\frac{Eta^2}{n^4} \frac{d^4 u_{zn}(x)}{dx^4} + D_b \left(\frac{n^2 - 1}{a^2}\right)^2 u_{zn}(x) = p_{zn}(x) \qquad (10.43)$$

which is the corresponding approximation of differential equation (10.21).

The above differential equation is similar to the one for a beam on an elastic foundation if the modulus of subgrade is taken as $D_b((n^2 - 1)/a^2)^2$ and the flexural rigidity of the beam is described by Eta^2/n^4. Hence, it is observed that the full circular cylinder under non-axisymmetric loading behaves as a curved membrane that is elastically supported by the so-called ring bending action.

Reference

1. Pircher M, Guggenberger W, Greiner R (2001) Stresses in elastically supported cylindrical shells under wind load and foundation settlement. Adv Struct Eng 4(3):159–167

Chapter 11
Analysis by Circular Cylindrical Super Elements

By nature, an analysis based on the theories in Chaps. 9 and 10 is very cumbersome. Therefore, we made a special computer program CShell available for circular cylindrical shells, in which we implemented the Donnell theory and Morley theory. This chapter is added for completeness. For readers who are interested only in the scope and options of the program, it is sufficient to read Sect. 11.1. They then may proceed to the next chapter without loss of essential information.

11.1 Introduction of Super Element Analysis

The program CShell is based on the stiffness method as applied in standard Finite Element Analysis software. We assume that the reader is familiar with the general set-up of such programs. The shell structure is divided into a number of elements, which are connected in nodes. Displacements in the nodes play the role of fundamental unknowns. Distributed loads can be applied over the area of the elements and lumped loads can be applied in the nodes. Standard commercially available software presupposes division of the surface of the cylindrical shell in elements in both circumferential direction and longitudinal direction. In general, the elements are small and the number of elements is large. For the special program, discussed in this chapter, we need not divide the shell in circumferential direction. Division in elements is done in longitudinal direction only. As a consequence, an element is a full circular cylindrical shell part, and the circular interface between two such parts is a *circular node*. We call a shell part a *super element*. The shell thickness is constant within a super element. At the nodes, we may add *ring elements (stiffeners)*. Figure 11.1 clarifies the composition of a shell from circular cylinders, circular nodes and circular rings. Loading may exist of their own weight, wind, and so on. The program is able to process loads p_x in longitudinal direction, p_z in radial direction and p_θ in circumferential direction. Per super element, the load is constant or linear in longitudinal direction. In circumferential direction, the only requirement is that the load is symmetric with respect to a chosen generating curve

J. Blaauwendraad and J. H. Hoefakker, *Structural Shell Analysis*, 173
Solid Mechanics and Its Applications 200, DOI: 10.1007/978-94-007-6701-0_11,
© Springer Science+Business Media Dordrecht 2014

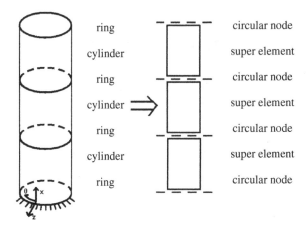

Fig. 11.1 Example of a shell structure modelled by super elements

(meridian). A super element edge will start or end at (i) a support or free edge, (ii) a change of shell thickness, (iii) a stiffening ring, (iv) a change of the intensity of the surface load, or (v) a circumferential line load. In case of a shell with a constant geometry and material properties, which is subject to linearly distributed surface loads and nodal line loads and/or point loads at the edge circles, only one super element is needed.

11.2 Outline of Super Element Analysis

We base the working out of the super element on the general solution for circular cylindrical shells in Chap. 7 for the Donnell theory (worked out in Table 9.2) and in Chap. 9 for the Morley theory (Table 9.1). We apply the same coordinate system x, θ, z, and use the same displacements u_x, u_θ, u_z and associated loads p_x, p_θ, p_z. The same holds for the strain and stress quantities. Also the sign conventions are the same. As said, we restrict the working out to symmetric loads.

11.2.1 Consideration of Super Element Level

We define the line of symmetry at $\theta = 0$, and observe that the loads p_x and p_z are even periodic functions with period $2\pi/n$ with respect to that line of symmetry, and that the load p_θ is an odd periodic function, where n is the mode number and represents the number of whole waves in circumferential direction. The Fourier series of any even or odd function consists only of the even trigonometric functions $\cos n\theta$ or odd trigonometric functions $\sin n\theta$, respectively, and a constant term. Therefore, the three load components can be described by a Fourier trigonometric series expressed by

$$p_x(\theta) = \sum_{n=0}^{\infty} p_{xn} \cos n\theta$$

$$p_\theta(\theta) = \sum_{n=0}^{\infty} p_{\theta n} \sin n\theta \tag{11.1}$$

$$p_z(\theta) = \sum_{n=0}^{\infty} p_{zn} \cos n\theta$$

In keeping with the trigonometric series load, a trial homogeneous solution u^h to the reduced differential equation will be of the trigonometric series form. Similarly, also the inhomogeneous solution u^i will have the same circumferential distribution. So, in correspondence with the distribution of the load components, the general solution for the displacements is of the congruent form

$$u_x(x, \theta) = \sum_{n=0}^{\infty} \left[C_h u_x^h(x) + u_x^i(x) \right] \cos n\theta$$

$$u_\theta(x, \theta) = \sum_{n=0}^{\infty} \left[C_h u_\theta^h(x) + u_\theta^i(x) \right] \sin n\theta \tag{11.2}$$

$$u_z(x, \theta) = \sum_{n=0}^{\infty} \left[C_h u_z^h(x) + u_z^i(x) \right] \cos n\theta$$

where C_h represents the eight arbitrary constants per circumferential mode number. The degrees of freedom in circular nodes are u_x, u_θ, u_z and φ_x, where φ_x is the rotation in x-direction of the circular edge of the super element. The rotation is positive if it raises a positive displacement u_x at the shell surface with a positive z-coordinate. The associated generalized edge forces are f_x, f_θ, f_z and t_x, where t_x is a moment per unit length. Also here, even quantities have a cosine series development and odd ones a sine series development. On the basis of the same consideration, we conclude that strain and stress quantities in the super element are described by functions of either the form $\cos n\theta$ or $\sin n\theta$.

An important conclusion is that the relation between quantities in the super element and associated quantities at the edge of the element only depend on the axial distribution. In other words, a stiffness relation for the super element depends on the amplitude of the circumferential distribution (which depends on the circumferential mode number n) but the trigonometric distribution needs not to be taken into account. Hence, for every possible mode number n the general solution for the degrees of freedom can be represented by

$$\begin{bmatrix} \hat{u}_x(x) \\ \hat{u}_\theta(x) \\ \hat{u}_z(x) \\ \hat{\phi}_x(x) \end{bmatrix} = \begin{bmatrix} A_{11}(x) & \cdots & A_{18}(x) \\ \vdots & \ddots & \vdots \\ A_{41}(x) & \cdots & A_{48}(x) \end{bmatrix} \begin{bmatrix} C_1 \\ \vdots \\ \vdots \\ C_8 \end{bmatrix} + \begin{bmatrix} \hat{u}_x^i(x) \\ \hat{u}_\theta^i(x) \\ \hat{u}_z^i(x) \\ \hat{\phi}_x^i(x) \end{bmatrix} \tag{11.3}$$

or briefly as

$$\hat{\mathbf{u}}(x)^c = \mathbf{A}(x)^c \mathbf{c} + \hat{\mathbf{u}}^i(x)^c \tag{11.4}$$

where $\mathbf{A}(x)^c$ is a rectangular matrix of size 4×8, of which the coefficients depend on the element geometry, the material properties and the mode number n. The hat notation refers to amplitude and the superscript c represents that the matrix equation refers to continuous quantities.

With the objective of formulating expressions for the edge forces in mind, the internal stress quantities have to be transformed into suitable quantities according to the boundary conditions formulated in Sect. 6.6 for the Donnell theory and Sect. 9.2.4 for the Morley theory. Performing the above-mentioned substitutions and transformation, we obtain the general solution for the internal stress quantities, which can be represented by

$$\begin{bmatrix} \hat{n}_{xx}(x) \\ \hat{n}_{x\theta}^*(x) \\ \hat{v}_x^*(x) \\ \hat{m}_x(x) \end{bmatrix} = \begin{bmatrix} B_{11}(x) & \cdots & B_{18}(x) \\ \vdots & \ddots & \vdots \\ B_{41}(x) & \cdots & B_{48}(x) \end{bmatrix} \begin{bmatrix} C_1 \\ \vdots \\ \vdots \\ C_8 \end{bmatrix} + \begin{bmatrix} \hat{n}_{xx}^i(x) \\ \hat{n}_{x\theta}^{*i}(x) \\ \hat{v}_x^{*i}(x) \\ \hat{m}_{xx}^i(x) \end{bmatrix} \tag{11.5}$$

or briefly as

$$\hat{\mathbf{n}}(x)^c = \mathbf{B}(x)^c \mathbf{c} + \hat{\mathbf{n}}^i(x)^c \tag{11.6}$$

where the matrix $\mathbf{B}(x)^c$ is also a rectangular matrix of size 4×8 of which the coefficients depend on the element geometry, the material properties and the mode number n.

We can easily transform the general solutions of Eqs. (11.3) and (11.5) in expressions for the edge displacements and edge forces of the element by substituting the nodal coordinates of the parallel edge lines. While formulating expressions for the edge forces, it is necessary to take into account that internal stress quantities on the negative side of the element act in negative coordinate direction and thus in opposite direction to the positive direction of the edge forces. Identifying one edge with $x = a$ and the other with $x = b$, the expressions for the element displacements read

$$
\begin{bmatrix} \hat{u}_x(a) \\ \hat{u}_\theta(a) \\ \hat{u}_z(a) \\ \hat{\varphi}_x(a) \\ \hat{u}_x(b) \\ \hat{u}_\theta(b) \\ \hat{u}_z(b) \\ \hat{\varphi}_x(b) \end{bmatrix} = \begin{bmatrix} A_{11}(a) & \cdots & A_{18}(a) \\ \vdots & \ddots & \vdots \\ A_{41}(a) & \cdots & A_{48}(a) \\ A_{11}(b) & \cdots & A_{18}(b) \\ \vdots & \ddots & \vdots \\ A_{41}(b) & \cdots & A_{48}(b) \end{bmatrix} \begin{bmatrix} C_1 \\ C_2 \\ C_3 \\ C_4 \\ C_5 \\ C_6 \\ C_7 \\ C_8 \end{bmatrix} + \begin{bmatrix} \hat{u}_x^i(a) \\ \hat{u}_\theta^i(a) \\ \hat{u}_z^i(a) \\ \hat{\varphi}_x^i(a) \\ \hat{u}_x^i(b) \\ \hat{u}_\theta^i(b) \\ \hat{u}_z^i(b) \\ \hat{\varphi}_x^i(b) \end{bmatrix} \tag{11.7}
$$

and the element forces read

$$
\begin{bmatrix} \hat{f}_x(a) \\ \hat{f}_\theta(a) \\ \hat{f}_z(a) \\ \hat{t}_x(a) \\ \hat{f}_x(b) \\ \hat{f}_\theta(b) \\ \hat{f}_z(b) \\ \hat{t}_x(b) \end{bmatrix} = \begin{bmatrix} B_{11}(a) & \cdots & B_{18}(a) \\ \vdots & \ddots & \vdots \\ B_{41}(a) & \cdots & B_{48}(a) \\ B_{11}(b) & \cdots & B_{18}(b) \\ \vdots & \ddots & \vdots \\ B_{41}(b) & \cdots & B_{48}(b) \end{bmatrix} \begin{bmatrix} C_1 \\ C_2 \\ C_3 \\ C_4 \\ C_5 \\ C_6 \\ C_7 \\ C_8 \end{bmatrix} + \begin{bmatrix} \hat{f}_x^i(a) \\ \hat{f}_\theta^i(a) \\ \hat{f}_z^i(a) \\ \hat{t}_x^i(a) \\ \hat{f}_x^i(b) \\ \hat{f}_\theta^i(b) \\ \hat{f}_z^i(b) \\ \hat{t}_x^i(b) \end{bmatrix} \tag{11.8}
$$

Briefly, these two equations become

$$
\hat{\mathbf{u}}^e = \mathbf{A}^e \mathbf{c} + \hat{\mathbf{u}}^{i;e} \tag{11.9}
$$

and

$$
\hat{\mathbf{f}}^e = \mathbf{B}^e \mathbf{c} + \hat{\mathbf{f}}^{i;e} \tag{11.10}
$$

where $\mathbf{A}^e$ and $\mathbf{B}^e$ are square matrices of size 8×8, of which the coefficients depend on the element geometry, the material properties and the mode number n. The hat notation refers to amplitude and the superscript e represents that the matrix equation refers to element quantities. The element stiffness matrix relates the element displacements in Eq. (11.9) to the element forces in Eq. (11.10). Therefore, the constants should be eliminated, which is done by first rearranging Eq. (11.9) to

$$
\mathbf{c} = \mathbf{A}^{-1;e} \left(\hat{\mathbf{u}}^e - \hat{\mathbf{u}}^{i;e} \right) \tag{11.11}
$$

and by substituting this expression into Eq. (11.10) resulting in

$$
\hat{\mathbf{f}}^e = \mathbf{B}^e \mathbf{A}^{-1;e} \left(\hat{\mathbf{u}}^e - \hat{\mathbf{u}}^{i;e} \right) + \hat{\mathbf{f}}^{i;e}. \tag{11.12}
$$

Case 1   We now consider two cases. In the first case, no distributed load is acting on the super element. Then $\hat{\mathbf{u}}^{i;e} = \hat{\mathbf{f}}^{i;e} = 0$. Then Eq. (11.12) reduces to

$$
\hat{\mathbf{f}}^e = \mathbf{K}^e \hat{\mathbf{u}}^e \tag{11.13}
$$

where the stiffness matrix $\mathbf{K}^e$, a square matrix of size 8×8, is

$$\mathbf{K}^e = \mathbf{B}^e \mathbf{A}^{-1;e} \tag{11.14}$$

Case 2 In the second case, the super element is loaded by a distributed load, while all edge displacements $\hat{\mathbf{u}}^e$ are zero. In this case, we call the element forces $\hat{\mathbf{f}}^{prime;e}$, and from Eq. (11.12) we derive

$$\hat{\mathbf{f}}^{prime;e} = -\mathbf{B}^e \mathbf{A}^{-1;e} \hat{\mathbf{u}}^{i;e} + \hat{\mathbf{f}}^{i;e} \tag{11.15}$$

The vector $\hat{\mathbf{f}}^{prime;e}$ comprises edge loads that are in equilibrium with their distributed load on the super element. In reverse, the load by the super element on the circular node is $-\hat{\mathbf{f}}^{prime;e}$.

11.2.2 Load on Circular Node

At a node, an external force $\hat{\mathbf{f}}^{ext;n}$ can be applied, where the superscript n refers to a nodal quantity. If more elements occur, two elements share one common node. The total load vector at a node $\hat{\mathbf{f}}^{tot;n}$ is given by the external load and the primary load vector of the adjacent super elements:

$$\hat{\mathbf{f}}^{tot;n} = \hat{\mathbf{f}}^{ext;n} - \sum_e \hat{\mathbf{f}}^{prim;e;n} \tag{11.16}$$

11.2.3 Assembling and Solving Procedure

The process of the assembly resulting in the global matrix equation, the incorporation of the prescribed displacements and the solution of the resulting reduced global matrix equation are to be done according to the common procedure of the stiffness method. Consequently, we can easily take care of, for example, an elastic support and stiffening ring elements or assembly of different geometries.

The solution of the resulting global matrix equation yields the magnitude of the nodal displacements and since these are equal to the element displacements $\hat{\mathbf{u}}^e$, the constants $\mathbf{c}$ per element can be computed by Eq. (11.11). Having obtained the constants, we can compute the continuous distribution of the displacements and stress quantities within the element by the expressions of Eq. (11.3) and Eq. (11.5), respectively. Finally, element forces and support forces can be determined from these solutions.

11.3 Calculation Scheme

We thus perform the following steps by a finite element program that is suited for super elements:

1. Read number of elements and nodes;
2. Read geometry and material properties of each element and nodal coordinates;
3. Read initial displacements;
4. Read distributed forces on the element and external forces on the nodes;
5. Compute matrices $\mathbf{A}^e$ and $\mathbf{B}^e$;
6. Generate the element stiffness matrix $\mathbf{K}^e$ according to Eq. (11.14);
7. Assemble the global stiffness matrix $\mathbf{K}$ via a location procedure;
8. Compute the primary load vector $\hat{\mathbf{f}}^{prim;e}$ per element according to Eq. (11.15);
9. Generate the load vector at a node $\hat{\mathbf{f}}^{tot;n}$ according to Eq. (11.16);
10. Assemble the global load vector $\hat{\mathbf{f}}^{tot}$ via a location procedure;
11. Compose the system of equations and incorporate the prescribed displacements;
12. Solve the system of equations to obtain the nodal displacements;
13. Compute per element the constants $\mathbf{c}$ according to Eq. (11.11);
14. Obtain the continuous distribution of the displacements and stress quantities of each element according to expressions similar to Eqs. (11.3) and (11.5), respectively;
15. Solve element and support forces.

A flow chart of such a program is given in Fig. 11.2.

11.4 Features of the Program CShell

In general, the program CShell facilitates analysis of structures as shown in Fig. 11.1, which consists of a number of cylindrical super elements, ring elements, circular nodes and supports. As explained before, if the shell to be analyzed has a constant geometry and material properties, and the load is distributed constantly or linearly over the height and/or consists of loads in the circular nodes, we need only one super element.

11.4.1 Structure, Supports and Loading

The program CShell can be applied to calculate the response of thin shell structures, that:

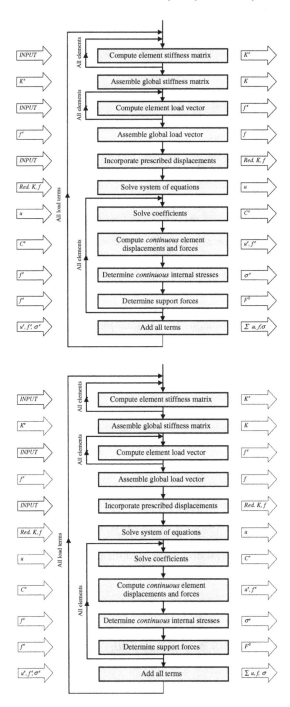

Fig. 11.2 Flow chart of the computer program with super elements

- Consist of circular cylindrical elements and ring stiffening elements, see Fig. 11.1; extension and bending of the stiffener is taken into account while torsion is neglected;
- Are subject to static distributed surface loads, circumferential line loads and point loads; these loads are symmetric with respect to the axis $\theta = 0$ and developed in trigonometric series in circumferential direction (in θ-direction, e.g., modes $p_n \cos(n\theta)$ for $n = 0, 1, 2, \ldots$); results of the calculation appear as one or all terms of the trigonometric series;
- Have constant linear elastic material properties per element;
- Have constant geometrical properties per element; and
- Are supported by fixed and/or elastic supports.

11.4.2 Shell Theory to be Chosen

Two theories for thin shells have been implemented, the theory of Morley and the theory of Donnell. The theory of Morley is considered to be the most exact one because it uses more appropriate kinematical relations for the changes of curvature. The Donnell theory is less accurate for lower modes and clearly fails for mode $n = 2$. On the other hand, to obtain the solution to the Morley theory an approximation was introduced which limits its applicability for higher modes. Therefore, Morley is used for lower modes and is compulsory for mode $n = 2$. For higher modes, Donnell gives more reliable results. The user indicates at which mode the switch between Morley and Donnell should be made. In case of default, Donnell is adopted for $n = 6$ and higher.

For the formulations resulting from the Morley equation that we implemented in the program CShell we refer to derivations in previous chapters. The stiffness matrix is based on the approximate expressions for the homogeneous solution presented in Sect. 9.4.2. We have used the inhomogeneous solutions to derive the load vectors. These are presented for mode number $n = 0$ in Chap. 5, for $n = 1$ in Chap. 4 as indicated in Sect. 9.9, and for $n \geq 2$ in Sect. 9.7.

The formulations resulting from the Donnell equation that we implemented in the program CShell are provided in Table 9.2.

11.4.3 Ring Elements

The ring analysis is largely based on the same set of relations as for a circular cylindrical shell on the basis of the Morley theory, with all quantities in axial direction being omitted. The result is thus a stiffness relation between the displacements of the ring element and the associated generalized line forces.

11.4.4 Verification

In verifying CShell by Finite Element Analysis, we obtained an excellent agreement with respect to displacement and deformed shape of the models. The axial stress and shear stress are accurately predicted with respect to magnitude and shape. The circumferential stresses (mainly the membrane component of those stresses) are less accurate, which is closely related to the simplifications introduced to arrive at the Morley equation (refer to Sect. 9.3.2). Only negligible numerical differences could be detected between the respective results. Based on these observations, the numerical capability of the developed program and the tremendous benefit of the super element approach for rational first-estimate design are conclusively demonstrated.

11.4.5 Output

The following automated output is available:

- Line plots in axial as well as circumferential direction.
- The deformation of a circular profile as well as the whole structure.
- A data file to select the quantities of interest and their respective location.

11.5 Overview of the Analysed Structures

In this book, the following structures are studied with the aid of the developed program:

- Chimneys, which are supported at the bottom, with or without stiffening rings and elastic supports (Chap. 12); and
- Tanks, which are supported at the bottom, with or without a roof or stiffening ring at the top and under full circumferential settlement (Chap. 13).

The chimneys are all loaded by a wind load. As described in Sect. 12.1, this wind load is developed into a quasi-static load series. The advantage is that each possible load-deformation behaviour (as described in Sect. 9.1.4) is present. Hence, the different response for the same geometry enables the interpretation and enlarges the understanding of the phenomena that occur per mode number.

The tanks are loaded by a content or wind load or subject to a full circumferential settlement. These cases represent the three main load-deformation conditions that can be identified for the overall response of the tank wall.

Chapter 12
Chimneys

In this chapter, we obtain closed-form solutions by adopting the formulations derived in Chaps. 9 and 10 for stiffened and non-stiffened long circular cylindrical shell structures and for various support conditions and loading terms. We will present design formulas including the range of application for chimneys upon comparison with solutions obtained by the computer program CShell of Chap. 11.

12.1 Wind Load

The distribution of the wind load around a circular cylindrical chimney has a maximal value at the windward meridian (denoted by $\theta = 0$) equal to the stagnation pressure p_w and a small pressure at the leeward meridian. The sides in between are subjected to suction, which in absolute value is even larger than the stagnation pressure (see Fig. 12.1 for a typical distribution at the left).

Because of the choice of the coordinate system and the symmetry of the load, we can develop the wind load (constant in axial direction) in a Fourier cosine series for the circumferential direction. By sign convention, we take the positive direction of the load in the positive direction of the coordinate z, which is from inside to outside of the circular profile. For a quasi-static load series, only the lower mode numbers have to be taken into account to accurately describe the wind load. For instance, the distribution exemplified in Fig. 12.1 may be defined by:

$$p_z(x, \theta) = p_w[\alpha_0 + \alpha_1 \cos \theta + \alpha_2 \cos 2\theta + \alpha_3 \cos 3\theta + \alpha_4 \cos 4\theta + \alpha_5 \cos 5\theta]$$

$$(12.1)$$

in which $\alpha_0 = 0.823$; $\alpha_1 = -0.448$; $\alpha_2 = -1.115$; $\alpha_3 = -0.400$; $\alpha_4 = 0.113$; $\alpha_5 = 0.027$; where p_w is set equal to 1 kN/mm^2, which value is a good reference value for the wind stagnation pressure in north-western Europe. The shape of the circumferential distribution of the wind load depends roughly on the geometry of

J. Blaauwendraad and J. H. Hoefakker, *Structural Shell Analysis*,
Solid Mechanics and Its Applications 200, DOI: 10.1007/978-94-007-6701-0_12,
© Springer Science+Business Media Dordrecht 2014

the chimney and varies from code to code but has the common characteristic that only a part of the circumference, the so-called stagnation zone, is under circumferential compression, while the remainder is under suction. The values presented above are representative for a long vertical cylinder and for short cylinders different values should be considered depending on the Reynolds number corresponding to the geometry of the cylinder.

12.2 Fixed Base: Free End

12.2.1 Closed-Form Solution

The circumferential distribution of the axial membrane stress resultant n_{xx} at the clamped base of a long circular cylinder (for example an industrial, steel chimney) under the wind load described in Sect. 12.2 has been subject to various studies. So far, we obtained closed-form solutions mainly with the aid of the Donnell equation, while this section presents the closed-form solution to the Morley equation. The axial stress distribution at the base of such a long chimney is mainly described by the beam action. As a result, σ_{beam} as depicted at the right in Fig. 12.1 occurs. However, the large suction at the sides of the chimney leads to an additional out of roundness of the cross-section, e.g., for $n = 2$ the circular cross-section deforms to an oval shape. To withstand this out-of-roundness at the base, additional axial stresses are generated. The stress increases to σ_{total} in Fig. 12.1. This is due to restrained warping. The chimney wants to ovalize, but can not at the base (denoted by $x = 0$). Then the cross-section will warp, however, at the base the warping is restricted, There the cross-section must remain plane, which causes additional membrane stresses in axial direction.

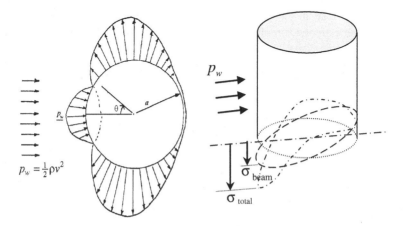

Fig. 12.1 Typical distribution of the wind load (*left*) and axial stress at the base (*right*)

At the base the chimney is typically clamped and at the top (denoted by $x = l$) the chimney often has a free edge. Over the distance l between these two edges, the geometrical and material properties are assumed to be constant. This means that the response to the wind load can be calculated by the solution to the differential equation (9.31). We must complement this solution by the appropriate boundary conditions:

$$x = 0; \quad \text{clamped:} \quad u_x = 0; \quad u_\theta = 0; \quad u_z = 0; \quad \varphi_x = 0$$
$$x = l; \quad \text{free:} \quad n_{xx} = 0; \quad n_{x\theta} = 0; \quad v_x^* = 0; \quad m_{xx} = 0$$
(12.2)

Response to Axisymmetric Load Term ($n = 0$)

The first term ($n = 0$) of the series development for the wind load in Eq. (12.1) is constant in circumferential direction and represents axisymmetric loading. It leads to a small circumferential tension in the chimney and due to the clamped edge to a short edge disturbance. However, the resulting stresses and displacements are known to be negligible in comparison with the response to the other terms of the wind load.

Response to Membrane Load Term ($n = 1$)

The second term ($n = 1$) describes a varying load that has a negative peak value at the windward meridian and a positive peak value at the leeward meridian. This is the only load term that is not self-balancing: i.e. it has a resultant in the wind direction. If the chimney is long, the stresses and deformations due to this load may be calculated by the membrane theory. However, not all boundary conditions of Eq. (12.2) can be fulfilled since there are more conditions than quantities, but the necessary edge disturbance will be represented by a small influence over a short length. The same result can be obtained by elementary beam theory if the shear deformation is accounted for. In fact, the solution to this term is well known and by solving the boundary conditions for the membrane stress resultants at $x = l$ the following expression for the axial membrane stress resultant n_{xx} is obtained, which is quadratic with respect to the axial coordinate

$$n_{xx}(x, \theta) = -\frac{p_w \alpha_1}{2a} (l - x)^2 \cos \theta. \tag{12.3}$$

However, if the more complete solution of Eq. (9.71) as described in Sect. 9.9.2 is employed, it is shown that the common assumption that the membrane solution is accurate is slightly in error if the lateral contraction is accounted for. Due to the then arising incompatibility at the clamped edge, a small but evident edge disturbance is produced and the resulting bending stress couple m_{xx} does contribute to a certain extent to the axial stress at the base. First solving the boundary conditions for the stress resultants at the free edge ($x = l$), we obtain four constants:

$$C_3 = 0; \quad C_4 = 0; \quad C_5 = -\frac{p_{z1} l a}{Et}; \quad C_6 = \frac{1}{2} \frac{p_{z1} l^2}{Et} \tag{12.4}$$

in which $p_{z1} = p_w \alpha_1$. We now solve the boundary conditions for the clamped edge $(x = 0)$, which for a long chimney $(l/a \gg 5)$ results in

$$C_1 = -\frac{v p_{z1} l^2}{2 \, Et}; \quad C_2 = -\frac{v p_{z1} l^2}{2 \, Et}; \quad C_7 = v \frac{p_{z1} l a}{Et}; \quad C_8 = \frac{2 + v}{2} \frac{p_{z1} l^2}{Et} \tag{12.5}$$

The solution for n_{xx} and m_{xx} is obtained by back substitution and by introducing $p_{z1} = p_w \alpha_1$ for the wind load. The solution at $x = 0$ reads

$$n_{xx}(0, \theta) = -\frac{p_w \alpha_1 l^2}{2a} \cos \theta; \quad m_{xx}(0, \theta) = -v \frac{p_w \alpha_1 l^2}{4 \beta^2 a^2} \cos \theta \tag{12.6}$$

Hence, the effect of the bending stress couple is mainly limited to the short influence length, but is certainly not negligible at the base of the chimney. The corresponding axial stress at the base $x = 0$ due to the "beam term", as obtained by Eq. (9.17), is

$$\sigma_{xx}^{n=1}(0, \theta, z) = \frac{n_{xx}}{t} + \frac{2z}{t} \frac{6 m_{xx}}{t^2} = -\frac{p_w \alpha_1 l^2}{2at} \left(1 + \frac{2z}{t} \sqrt{3} \frac{v}{\sqrt{1 - v^2}} \right) \cos \theta \tag{12.7}$$

Note that the load factor α_1 is negative for the current wind load. Hence, the axial stress is positive (tension) at the windward meridian $(\theta = 0)$ and negative (compression) at the leeward meridian $(\theta = \pi)$. The additional bending stress is only present over a short influence length, however, the contribution can be quite substantial. For the outer or inner surface $(z = \pm t/2)$ of, e.g., steel with $v = 0.3$, the term between the brackets becomes 1 ± 0.5. Hence, a prediction by the membrane stress resultants only, might be in a rather large error for such a material (in this case an error of 50 %).

Response to Self-balancing Load Terms ($n = 2$ and Larger)

The third term ($n = 2$) describes a double symmetric and hence self-balancing term with two waves about the circumference, which results in a pressure at the windward and the leeward meridian and a suction at the sides. We calculate the response to this load by using the homogeneous solution as presented in Table 9.1, which is complemented by the boundary conditions at hand.

The higher-order terms of the series development $n > 2$ are also self-balancing and therefore analysed with the same solution procedure as for $n = 2$, however, with their respective value of the circumferential wave number.

As shown in Sect. 9.7, we can obtain the inhomogeneous solution omitting all derivatives with respect to the axial coordinate x. For the present load

$$p_z(x, \theta) = \sum_n p_{zn} \cos n\theta \tag{12.8}$$

an inhomogeneous solution for $(n \geq 2)$ reads

$$u_z(x,\theta) = u_{zn}\cos n\theta = \frac{1}{D_b}\sum_{n=2}^{\infty}\frac{a^4}{(n^2-1)^2}p_{zn}\cos n\theta$$

$$u_\theta(x,\theta) = u_{\theta n}\sin n\theta = -\frac{1}{n}u_{zn}\sin n\theta$$

$$m_{\theta\theta}(x,\theta) = m_{\theta\theta n}\cos n\theta = \sum_{n=2}^{\infty}\frac{a^2}{n^2-1}p_{zn}\cos n\theta \qquad (12.9)$$

$$m_{xx}(x,\theta) = m_{xxn}\cos n\theta = \upsilon m_{\theta\theta n}\cos n\theta$$

$$v_\theta(x,\theta) = v_{\theta n}\sin n\theta = -\sum_{n=2}^{\infty}\frac{na}{n^2-1}p_{zn}\sin n\theta$$

where the other stress quantities are equal to zero. Apparently, the inhomogeneous solution for $n \geq 2$ is the ring-bending solution. It shows that the displacements u_z and u_θ are not equal to zero. The boundary conditions at the clamped edge $(x=0)$ are therefore not fulfilled and an edge disturbance that originates from this edge is necessary. The resulting edge disturbance has a far-reaching influence; the long influence length plays a role. The boundary condition for m_{xx} at $x=l$ is also not fulfilled but only due to a non-zero change of curvature in circumferential direction that is multiplied by Poisson's ratio v. We conclude that this fact alone leads to a short edge disturbance that originates from this free edge with a mainly local effect and a small influence on the response of the cylinder.

From the above mentioned arguments, it can be concluded that for a chimney with a length larger than the long influence length only the boundary conditions at the base are necessary to describe the overall response to the wind load. Hence, the constants in the homogeneous solution of the edge disturbance that originates from the free edge can safely be equated to zero. The expressions for the four quantities, which have to be described at the clamped edge, are listed in Table 9.1. We can now formulate the boundary conditions for this edge by adding the inhomogeneous solution to the expressions for the homogeneous solution at $x=0$, which gives four equations with four unknown constants. Making use of the fact that terms multiplied by $(a\beta)^{-4}$ are negligibly small in comparison to unity (for the lower values of n under consideration), we obtain the solution to these equations for $v=0$

$$C_1^n = 0; \quad C_2^n = 0; \quad C_5^n = -\hat{u}_z^n; \quad C_6^n = -\left(1 - \frac{n^2 - \frac{3}{2}}{(a\beta)^2}\right)\hat{u}_z^n \qquad (12.10)$$

where $\hat{u}_z^n$ is equal to u_{zn} as presented in the inhomogeneous solution above. For the case that Poisson's ratio is not zero, the solution is

$$C_1^n = C_2^n = -\frac{v}{2}\frac{n^2 - 1}{(a\beta)^2}\hat{u}_z^n; \quad C_5^n = -\left(1 - \frac{v}{2}\frac{n^2 - 1}{(a\beta)^2}\right)\hat{u}_z^n$$

$$C_6^n = -\left[1 - \left(n^2 - \frac{3}{2}\right)\frac{1}{(a\beta)^2} - v\left[\frac{3}{2}(n^2 - 1) - vn\sqrt{n^2 - 1}\right]\frac{1}{(a\beta)^2}\right]\hat{u}_z^n$$

$$(12.11)$$

The constants C_1^n and C_2^n represent the short-wave solution; they are zero for zero Poisson's ratio. Additionally, it can be verified that the long-wave solution (represented by the constants C_5^n and C_6^n) is mainly described by membrane stress resultants in the axial direction while the loading leads to bending stress resultants in circumferential direction.

For the free edge at $x = l$, we can apply a similar procedure to obtain the other four constants. As described, the boundary conditions at this edge are only not met by a bending stress couple, which occurs if the lateral contraction, described by Poisson's ratio v, is taken into account. For convenience, the solution is obtained at an edge $x = 0$ to cancel out the length in the expressions. Solving the four equations for the boundary conditions, we find the four constants

$$C_1^n = \frac{v}{2}\frac{n^2 - 1}{(a\beta)^2}\hat{u}_z^n; \quad C_2^n = -\frac{v}{2}\frac{n^2 - 1}{(a\beta)^2}\hat{u}_z^n; \quad C_5^n = \frac{v}{2}\frac{n^2}{(a\beta)^2}\hat{u}_z^n; \quad C_6^n = -\frac{v}{2}\frac{n^2}{(a\beta)^2}\hat{u}_z^n$$

$$(12.12)$$

which indeed shows that the long-wave solution is hardly activated since these constants are of the order $O\left(v(a\beta)^{-2}\right)$. The fact that the inhomogeneous solution is incompatible with the boundary conditions for the free edge is compensated by an edge disturbance that is described by a small short-wave and equally small long-wave solution.

On basis of these observations, it is obvious that the influence of the incompatibility at the free edge is negligible when calculating any quantity at the base of a sufficiently long cylinder. Additionally, the influence of the moment m_{xx} at the base on the axial stress distribution at the base is not negligible if the lateral contraction is accounted for. However, similar to the stress distribution for the "beam action", the contribution can be added to the membrane force n_{xx}. Moreover, the addition of the effects gives an identical ratio of the bending stress to the membrane stress. The expressions for the membrane force n_{xx} and the moment m_{xx} are found by back substitution of the homogeneous solution. Substitution of the constants and addition of the inhomogeneous solution results in the expressions of Eq. (12.13) in which the small terms of the order $O\left((a\beta)^{-2}\right)$ are neglected compared to unity.

$$n_{xx}(0, \theta) = -\sum_{n=2}^{5} 2\sqrt{3(1-v^2)} \frac{a^2}{t} \frac{p_{zn}}{n^2-1} \cos n\theta$$

$$m_{xx}(0, \theta) = -\sum_{n=2}^{5} va^2 \frac{p_{zn}}{n^2-1} \cos n\theta \tag{12.13}$$

Hence, we finally obtain the expression for the axial stress $\sigma_{xx}(x, \theta, z)$ at the base $x = 0$ due to the terms $(n \geq 2)$ by addition of the membrane and bending stress:

$$\sigma_{xx}^{2 \leq n \leq 5}(0, \theta, z) = \frac{n_{xx}}{t} + \frac{2z}{t} \frac{6m_{xx}}{t^2}$$

$$= -\sum_{n=2}^{5} 2\sqrt{3(1-v^2)} \frac{a^2}{t^2} \frac{p_{zn}}{n^2-1} \left(1 + \frac{2z}{t}\sqrt{3}\frac{v}{\sqrt{1-v^2}}\right) \cos n\theta$$

$$\tag{12.14}$$

We can derive a similar expression for the stress distribution at the base and at the middle surface $(z = 0)$ on the basis of the Donnell equation:

$$\sigma_{xx}^{2 \leq n \leq 5}(0, \theta, 0) = -\sum_{n=2}^{5} 2\sqrt{3(1-v^2)} \frac{a^2}{t^2} \frac{p_{zn}}{n^2} \cos n\theta. \tag{12.15}$$

The difference with the Morley solution is in the inhomogeneous part that describes the ring-bending action. It is well known that the Morley equation more accurately describes this part of the solution (especially for the case $n = 2$). The ratio of the solutions is equal to

$$\frac{\sigma_{xx}^{MK}(0, \theta, 0)}{\sigma_{xx}^{D}(0, \theta.0)} \approx \frac{n^2}{n^2-1} \quad \text{(for } 2 \leq n \leq 5). \tag{12.16}$$

Stress Distribution at the Base

Having found the response of the long chimney to the separate terms of the wind load, we can derive a useful design formula for the stress distribution at the base. For a chimney longer than the long influence length, it is readily verified that the only non-balancing term $(n = 1)$ is the leading term of the full response and conveniently, we can most easily find its response by a membrane solution or beam analysis. The other contributing terms are the self-balancing terms $(n = 2, \ldots, 5)$. The response of Eq. (12.14) to these load terms requires a more laborious solution and therefore it is convenient to express their influence by their ratio to the response to the 'beam term', as done in Eq. (12.7). This results in an expression for the axial stress at the base $x = 0$:

$$\sigma_{xx}^{0 \leq n \leq 5}(0, \theta, z) = \sigma_{xx}^{n=1}(0, \theta, z) \left[1 + \frac{\sum\limits_{n=2}^{5} \sigma_{xx}^{2 \leq n \leq 5}(0, \theta, z)}{\sigma_{xx}^{n=1}(0, \theta, z)} \right] \qquad (12.17)$$

Since the ratio of the bending-to-membrane stress for the non-balancing terms and the "beam term" are multiplied by the same factor, we can further simplify the formula to

$$\sigma_{xx}^{0 \leq n \leq 5}(0, \theta, \pm t/2) = \sigma_{xx}^{n=1}(0, \theta, 0) \left[1 + \frac{\sum\limits_{n=2}^{5} \sigma_{xx}^{2 \leq n \leq 5}(0, \theta, 0)}{\sigma_{xx}^{n=1}(0, \theta, 0)} \right]$$
$$\cdot \left(1 \pm \sqrt{3} \frac{v}{\sqrt{1 - v^2}} \right) \qquad (12.18)$$

The maximal tensile stress at $\theta = 0$ (the windward meridian) is

$$\sigma_{xx,t}^{0 \leq n \leq 5}(z = t/2) = -p_w \alpha_1 \frac{l^2}{2at} \left[1 + 4\sqrt{3(1 - v^2)} \left(\frac{a}{l} \right)^2 \frac{a}{t} \sum_{n=2}^{5} \frac{1}{n^2 - 1} \frac{\alpha_n}{\alpha_1} \right]$$
$$\cdot \left(1 + \sqrt{3} \frac{v}{\sqrt{1 - v^2}} \right) \qquad (12.19)$$

We obtain the formula for the maximal tensile stress at the clamped edge at the location of the windward meridian ($\theta = 0$) by substituting in this expression the wind load of Eq. (12.1):

$$\sigma_{xx,t}^{0 \leq n \leq 5}(z = t/2) = 0.224 \frac{l^2}{at} p_w \left[1 + 6.39\sqrt{1 - v^2} \left(\frac{a}{l} \right)^2 \frac{a}{t} \right]$$
$$\cdot \left(1 + \sqrt{3} \frac{v}{\sqrt{1 - v^2}} \right) \qquad (12.20)$$

The part of the expression in front of the straight brackets is the beam solution. The term between the straight brackets is the multiplier due to restrained warping of the cross-section at the base. The formula for the compressive stress at the middle surface ($z = 0$) and the leeward meridian ($\theta = \pi$) reads

$$\sigma_{xx,c}^{0 \leq n \leq 5}(z = 0) = -0.224 \frac{l^2}{at} p_w \left[1 - 4.88\sqrt{1 - v^2} \left(\frac{a}{l} \right)^2 \frac{a}{t} \right] \qquad (12.21)$$

which does not necessarily indicate the maximal compressive stress. The location of this maximum depends on the dimensions of the cylindrical shell and on the constants in the wind load.

Between the straight brackets of Eq. (12.20) for the tensile stress we recognize the inverse dimensionless parameters l/a and t/a. Note that, to obtain the stress on the outer or inner surface, the term between the round brackets has additionally to be taken into account. The practical range for long chimneys extends up to a value of around $l/a = 60$, which further depends on the thickness of the cylinder that may range up to a value of around $a/t = 400$.

The term between the round brackets depends only on the value of Poisson's ratio. The term between the straight brackets consists of the dimensionless parameters l/a and t/a which are multiplied by a factor. This factor depends on the constants in the series development of the wind load and is given by 6.39 if $v = 0$. The similar formula based on the solution to the Donnell yields 4.87 for that factor with $v = 0$. It shows that the Morley equation gives a tremendous improvement over the Donnell equation. Finite element analyses with nonzero Poisson's ratio support the high accuracy of the Morley solution.

We can present Eq. (12.20) in an alternative way. For this purpose, we introduce the characteristic lengths l_{c1} and l_{c2}:

$$l_{c1} = \sqrt{at}$$
$$l_{c2} = \sqrt[4]{atl^2} \tag{12.22}$$

Now we rewrite the formula for the maximal tensile stress at the clamped edge:

$$\sigma_{xx,t}^{0 \leq n \leq 5}(z = t/2) = 0.224 p_w \left(\frac{l}{l_{c1}}\right)^2 \left[1 + 6.39\sqrt{1 - v^2}\left(\frac{a}{l_{c2}}\right)^4\right]$$
$$\cdot \left(1 + \sqrt{3}\frac{v}{\sqrt{1 - v^2}}\right) \tag{12.23}$$

Note that the beam solution in front of the straight brackets depends on the characteristic length l_{c1}, and the multiplier term due to restrained warping between the straight brackets is dependent on the characteristic length l_{c2}. The multiplier term is plotted in Fig. 12.2 against the dimensionless parameter l_{c2}/a. For the practical range, the dimensionless ratios in Eq. (12.20) are $10 < l/a < 60$ and $50 < a/t < 400$, i.e. $0.7 < l_{c2}/a < 3$ and $70 < l/l_{c1} < 1200$ in Eq. (12.23). Note that the largest value for l_{c2}/a is obtained for the thickest and longest chimneys, i.e. smallest a/t in combination with largest l/a. Figure 12.2 shows that only for a considerable value of l_{c2}/a is the stress at the base of the chimney dominated by the beam behaviour. For shorter chimneys, the stress at the base varies not merely quadratically with the length as we might expect on the basis of beam theory, but is largely dominated by the non-balancing terms.

Hinged Versus Clamped Support

Until now we have obtained the solutions for a fully rigid support at the base of the chimney, i.e. a clamped edge. If the support of the chimney allows free rotation, the moment should be zero at the base. The solution for such a "hinged-wall" edge ($u_x = u_\theta = u_z = 0$, $m_{xx} = 0$) is almost equal to the solution for the clamped edge.

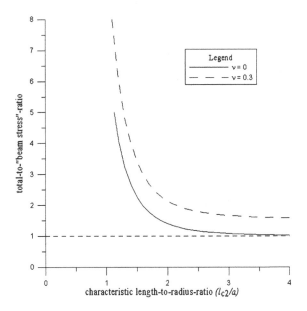

Fig. 12.2   Multiplier to beam solution

It is left to the reader to demonstrate that the change in the long edge disturbance is negligible, but the short edge disturbance is somewhat different. However, this difference is not of any importance with respect to the global solution for the stresses at the edge.

The tensile membrane stress at the base of a long chimney having either a clamped edge or a "hinged-wall" edge can thus be obtained by equating the product of the 'beam theory stress' with the multiplier for this 'beam theory stress' presented within the straight brackets of Eqs. (12.20) and (12.23). To obtain the maximum tensile stress for the clamped edged, we must additionally take into account the term between the round brackets of Eqs. (12.20) and (12.23), which only depends on the value of Poisson's ratio as shown in Fig. 12.2.

Lessons Learned

In this subsection, the solution to the Morley equation is used to obtain a suitable formula for the stress distribution at the fixed base of a long chimney under wind loading. Mainly because the inhomogeneous solution for the self-balancing terms $(n \geq 2)$ accurately describes the ring-bending action of the cylinder, the result is a substantial refinement of the formula on the basis of the Donnell equation and shows better agreement with finite element results.

The ratio of the total membrane stress to the 'beam theory stress' depends completely on the geometry of the chimney, the circumferential distribution of the wind load and to a lesser extent on the lateral contraction of the material. The influence of the additional stress, due to the higher-order terms of the wind loads, manifests itself in a long-wave solution. The shell behaviour in the part of the cylinder where the long-wave solution does not exert influence is in accordance with the ring-bending action. The long-wave solution represents the additional membrane action of the shell to meet the boundary conditions.

For the bending stress at the base of the circular cylinder, we have shown that a considerable contribution must be incorporated in the maximal tensile and compressive stress at the base. For steel with Poisson's ratio equal to $v = 0.3$, the bending stress is approximately 50 % of the membrane stress. This rather large increase is subdivided into two approximately equal parts: a part that produces the short edge disturbance and a part that produces a long edge disturbance.

It is noted that the result is obtained under the assumption that the length of the chimney is at least larger than the long influence length. For shorter cylinders, the solution cannot be obtained solely on the boundary conditions at the clamped base, since the long edge disturbance will produce stresses that are incompatible with the boundary conditions at the free end. Hence, a compensating long edge disturbance will originate from the free edge that might be of influence to the axial stress distribution. The range of application of the derived formula is the subject of the next subsection.

12.2.2 Applicability Range of Formulas

The formulas in Eqs. (12.20) and (12.23) predict the tensile axial stress at the base and the windward side of a long clamped chimney subject to wind load and only differ in the different dimensionless parameters that are adopted. The formulas describe the stresses at the middle surface and at the outer surface, respectively.

The range of application of these formulas can be investigated by comparison with results by the program CShell, which applies for both short and long cylindrical shells. For chimneys much longer than the influence length, we will obtain identical results. For chimneys shorter than the influence length, the program is more accurate since the formulas do not include the effect of the edge disturbance that originates at the free edge.

As an example, the multiplication factor for the outer fibre stress is plotted for the formula and for the CShell program in Fig. 12.3. The factor in this figure is thus based on all terms of Eq. (12.23). The agreement between the plots by the formula and the program is extremely good if the dimensionless parameter l_{c2}/a is larger than or equal to unity.

To interpret the dependency on this parameter, it is recalled that, besides the dependency on the wind load factors, the increase of the beam stress is attributed to the long-wave solution of the self-balancing terms of the wind load $(n \geq 2)$. Alternatively, we may state that the range of application of Eqs. (12.20) and (12.23) is correct for a length larger than the half influence length for $n = 2$ (the mode number that has the longest influence length and dominates that difference

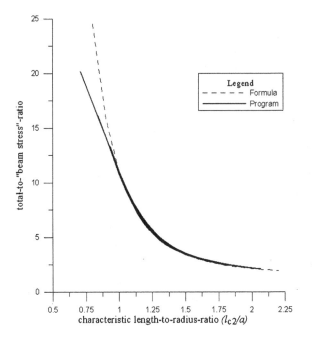

Fig. 12.3 Stress ratio for $v = 0.3$ at the outer fibre obtained by Eq. (12.23) and the program

between the beam stress and the total stress). For the cylinder equal to the half influence length, the edge disturbance starting from the base is reduced by a factor of the order $e^{-\pi/2}$ at the free edge. 'Reflection' from the free edge to the base yields this reduction factor again, so the disturbance is reduced to a value in the order of $e^{-\pi} \approx 0.043$. For a cylinder shorter than the half influence length, the resting value after reflection exponentially increases, which conclusively explains the difference between the two graphs in Fig. 12.3.

12.3 SMC Approximation

Similar to Sect. 12.2.1, we can derive a useful design formula for the stress distribution at the base of the long chimney by calculating the response to the wind load while adopting the SMC approximation as detailed in Chap. 10. Fully in line with the approach for the long chimney without stiffening rings, we can express the contribution of the self-balancing terms $(n = 2, \ldots, 5)$ by their ratio to the response to the "beam term" in Eq. (12.7). In order to obtain this ratio, we need first to solve the appropriate boundary conditions for the self-balancing terms as described in the SMC solution per Eq. (10.9). These are

$$x = 0; \quad \text{clamped:} \qquad u_x = 0; \quad u_\theta = 0$$
$$x = l; \quad \text{free:} \qquad n_{xx} = 0; \quad n_{x\theta} = 0 \tag{12.24}$$

If the chimney is long enough that the edge disturbance originating from the free edge does not influence the disturbance at the clamped edge, the solution for the constants becomes

$$C_1^n = -\hat{u}_z^n; \quad C_2^n = -\left(1 - \frac{3}{2}(1+v)\frac{n^2-1}{(\alpha\beta)^2}\right)\hat{u}_z^n; \quad C_3^n = C_4^n = 0 \tag{12.25}$$

where for the sake of comparison, the contributions of the order $(\alpha\beta)^{-2}$ are retained.

Substitution of the constants into the expression for the membrane force n_{xx} as presented in Table 10.1, results in the expression for n_{xx} at the clamped edge

$$n_{xx}(0, \theta) = -\sum_{n=2}^{5} 2\sqrt{3(1-v^2)}\frac{a^2}{t}\frac{p_{zn}}{n^2-1}\cos n\theta. \tag{12.26}$$

In this expression terms $O\left[(\alpha\beta)^2\right]$ are neglected compared to unity. The result is identical to the expression in Eq. (12.13) on the basis of the Morley equation. Having found the response for these self-balancing terms ($n = 2, \ldots, 5$), we can derive the design formula for the stress distribution at the base:

$$\sigma_{xx}^{0 \le n \le 5}(0, \theta, z) = \sigma_{xx}^{n=1}(0, \theta, z)\left[1 + \frac{\sigma_{xx}^{2 \le n \le 5}(0, \theta, z)}{\sigma_{xx}^{n=1}(0, \theta, z)}\right]. \tag{12.27}$$

We can assume that the ratio of the bending-to-membrane stress for the non-balancing terms is not altered, although not described by the SMC solution, and remains identical to ratio as obtained for the "beam term". Hence, we tentatively propose to further simplify the formula for the stress distribution at the base to

$$\sigma_{xx}^{0 \le n \le 5}(0, \theta, \pm t/2) = \sigma_{xx}^{n=1}(0, \theta, 0)\left[1 + \frac{\sigma_{xx}^{2 \le n \le 5}(0, \theta, 0)}{\sigma_{xx}^{n=1}(0, \theta, 0)}\right]$$
$$\cdot \left(1 \pm \sqrt{3}\frac{v}{\sqrt{1-v^2}}\right). \tag{12.28}$$

The formula for the maximal tensile stress at the clamped edge at the windward meridian ($\theta = 0$) for the long chimney based on the Morley equation is presented in Eq. (12.20) and this formula is obtained by substituting the wind load of Eq. (12.1). Performing the same substitutions for the SMC solution, the maximal tensile stress becomes

$$\sigma_{xx,t}^{0 \le n \le 5}(z = t/2) = 0.224 \frac{l^2}{at} p_w \left[1 + 6.39\sqrt{1 - v^2} \left(\frac{a}{l} \right)^2 \frac{a}{t} \right]$$
$$\cdot \left(1 + \sqrt{3} \frac{v}{\sqrt{1 - v^2}} \right),$$

$$(12.29)$$

which is identical to Eq. (12.20). This comparison with the Morley solution shows that a solution based on the SMC-equation is as accurate for the response of the long chimney to the self-balancing terms $(n = 2, \ldots, 5)$.

12.4 Effect of Ring Stiffeners

This section describes the influence of stiffening rings on the behaviour of the long chimney under wind load on basis of closed-form solutions and their range of application.

In Sects. 12.2 and 12.3, we have shown that the stress at the fixed base of a long cylinder under wind load can be conveniently related to the beam mode $(n = 1)$. The deformation and stress for the axisymmetric mode $(n = 0)$ are of no importance for the overall behaviour. Only the response to the higher modes $(n \geq 2)$ is markedly altered by the presence of stiffening rings in comparison with the response of a cylinder without rings. For these higher modes, the membrane force n_{xx} at the fixed base is directly related to the induced out-of-roundness (ovalisation) of the cylinder, which cannot occur at the base. As stated before, the cross-section tends to warp at the base. The normal stresses are needed to withstand this warping, in other words: to keep this section plane. As the presence of stiffening rings will reduce the ovalisation and hence the warping, n_{xx} is reduced accordingly, which is exactly what the SMC as presented in Sect. 12.4 describes.

12.4.1 Closed-Form Solution (SMC)

The semi-membrane concept (SMC), as presented in Chap. 10, is applicable to non-axisymmetric load cases of circular cylindrical shells provided that the cylinder is sufficiently long in comparison to its radius and that the boundary effects mainly influence the more distant material.

In the SMC approach, the bending stiffness of the shell is adopted only for the circumferential moment. As the ring behaviour can be adequately described by the bending action of the ring only, it is proposed to "smear out" the bending stiffness of the rings along the bending stiffness of the cylinder. This results in the following modified bending stiffness

$$D_{b,\mathrm{mod}} = D_b + \frac{EI_r}{l_r} \qquad (12.30)$$

where l_r denotes the spacing between the rings and I_r represents the bending stiffness of the ring. Hence, the difference between the solution for a long cylinder with multiple equidistant stiffening rings and the solution for a long cylinder without these rings can be captured by a modified parameter β_{mod} only. The definition of β in Eq. (7.18) changes into

$$\beta_{mod}^4 = \frac{D_m(1 - v^2)}{D_{b,mod} a^2}$$

(12.31)

It is convenient to define a dimensionless parameter λ_r as the ratio of the shell bending stiffness without and with stiffeners:

$$\lambda_r = \frac{D_b}{D_{b,mod}}$$

(12.32)

This facilitates to express β_{mod} into β:

$$\beta_{mod} = \beta \sqrt[4]{\lambda_r}$$

(12.33)

For the special case of rectangular stiffening rings with width b and height h located at the middle surface of the circular cylindrical shell, the stiffness ratio λ_r becomes

$$\lambda_r = \frac{l_r t^3}{l_r t^3 + bh^3(1 - v^2)}$$

(12.34)

Identical to Sect. 12.3, a useful design formula can be derived for the stress distribution at the base of the long chimney stiffened by equidistant rings by calculating the response to the wind load. With the same boundary conditions of Eq. (12.24), the solution for the constants becomes

$$C_1^n = -\hat{u}_z^n; \quad C_2^n = -\left(1 - \frac{3}{2}(1 + v)\frac{n^2 - 1}{(a\beta_{mod})^2}\right)\hat{u}_z^n; \quad C_3^n = C_4^n = 0 \quad (12.35)$$

where for the sake of comparison, the contributions of the order $(a\beta_{mod})^{-2}$ are retained.

Substitution of these constants into the expression for the membrane force n_{xx} as presented in Table 10.1, replacing β by β_{mod} and neglecting terms of $O\left[(\alpha\beta)^2\right]$ to unity, results in the expression for n_{xx} at the clamped edge

$$n_{xx}(0, \theta) = -\sum_{n=2}^{5} 2\sqrt{3(1 - v^2)}\frac{a^2}{t}\sqrt{\lambda_r}\frac{p_{zn}}{n^2 - 1}\cos n\theta$$

(12.36)

The resulting formula for the maximal tensile stress at the clamped edge at the windward meridian ($\theta = 0$) for the long chimney becomes

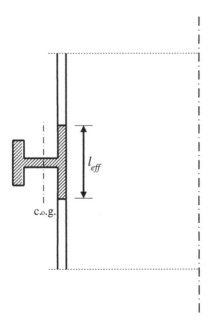

Fig. 12.4 Definition of eccentric ring

$$\sigma_{xx,t}^{0 \le n \le 5}(z = t/2) = 0.224 \frac{l^2}{at} p_w \left[1 + 6.39\sqrt{1 - v^2}\sqrt{\lambda_r}\left(\frac{a}{l}\right)^2\frac{a}{t}\right]$$
$$\cdot \left(1 + \sqrt{3}\frac{v}{\sqrt{1 - v^2}}\right) \tag{12.37}$$

in which the only change compared with Eq. (12.20) is the addition of the factor $\sqrt{\lambda_r}$ within the straight brackets.

The above solution has been derived for stiffening rings with their centre of gravity located at the middle surface of the cylinder, i.e. symmetric rings. For eccentric rings, the flexural rigidity of the combined ring and a certain effective length of cylinder should be adopted, which is to be evaluated with respect of the centre of gravity of this combined ring as depicted in Fig. 12.4. For engineering purposes, it is sufficient to determine the flexural rigidity as we do for a straight bar (i.e. neglecting the influence of the curvature).

The effective length is known to be a function of the shell radius, thickness wave number of the loading, the stiffener spacing and the ring dimensions and eccentricity. A proposal for the appropriate effective length to be accounted for within the present approach is provided in Sect. 12.5.2.

12.4.2 Applicability Range of Formulas

Similar to Sects. 12.2 and 12.3, the Eq. (12.37) predicts the tensile axial stress at the base and the windward side of a long clamped chimney subject to wind load. However, the influence of distributed stiffening rings is incorporated into the formula.

We have investigated the range of application of this formula by comparison with results obtained by the developed program, which applies for short and long cylindrical shells and allows accurate modelling of stiffening rings. For a chimney with closely spaced stiffening rings, the formula predicts an accurate value of the stress at the base, as this program is based on the closed-form solution and the formula is obtained by "smearing out" the bending stiffness of the rings. For chimneys shorter than, say, the influence length and/or for chimneys with a more uneven distribution of the ring stiffness, the program is more accurate than the formula since the formula does not include the effect of the edge disturbance that originates at the free edge and as the formula is based on a constant distribution of the ring stiffness along the length of the chimney.

Calculations have been performed for both stiffening rings with their centre of gravity located at the middle surface of the cylinder and for eccentric stiffening rings, of which the approach and results are discussed below.

Stiffening Rings. Centre of Gravity at the Middle Surface of the Cylinder

For stiffening rings with their centre of gravity located at the middle surface of the cylinder, the design formula has been verified with the developed program. Calculations have been made for a radius-to-thickness-ratio of 100, with length-to-radius-ratios of 10, 20 and 30 and with 2, 3, 4 and 5 equally spaced symmetric stiffening rings per length-to radius ratio. The stress ratio between the stress due to the mode numbers $n = 1$ and $n = 2$ and the stress due to the "beam term" has been obtained by the program. The calculated stress ratio is fairly in line with the stress ratio predicted by formula (12.37) unless the spacing between the stiffening rings is chosen too large. For stiffening rings with a spacing roundabout equal to and larger than half of the long influence length, the difference between the program results and the values predicted by the formula increase with increasing ring stiffness, i.e. decreasing stiffness ratio λ_r. The difference between the values predicted by the formula and the values obtained by the program is small for the cases with closely spaced stiffening rings, i.e. with a spacing shorter than half of the long influence length.

Eccentric Stiffening Rings to the Middle Surface of the Cylinder

For eccentric stiffening rings, the envisaged necessity to account for a certain effective shell length to determine the equivalent ring stiffness within the SMC approach has been confirmed by the reported program results. Calculations have been made for a radius-to-thickness-ratio of 100 and 200 and with 3 and 5 equally spaced stiffening rings per length-to-radius-ratio. For the radius-to-thickness-ratio of 100, the length-to-radius-ratios of 10, 20 and 30 have been considered and for the radius-to-thickness-ratio of 200, the length-to-radius-ratios of 15, 30 and 45 have been considered. For both radius-to-thickness-ratios, these respective length-to-radius-ratios approximately match with a 0.5, 1 and 1.5 times the influence length of the long-wave solution. Based on these program results, it has been shown that the determined effective lengths are (much) shorter than the existing formulation for the effective shell length, i.e. $1.56\sqrt{at}$. Furthermore, it has been shown that the effective shell length to be accounted for also depends on the stiffener spacing, ring dimensions and eccentricity.

To match with the results of the program, a preliminary proposal for the effective length has been provided based on the observations above. As a conclusive result could not be obtained, it is proposed to conservatively take the effective shell length equal to half of the existing formulation. Considering the applicability of the design formula, a marked improvement is already achieved by inclusion of a certain effective length and the need for more improvement within the practical ranges is considered to be unnecessary for rational first-estimate design of ring-stiffened circular cylindrical shells.

12.5 Effect of Elastic Supports

This section describes the influence of elastic supports on the behaviour of the long chimney under wind load. The influence is calculated on the basis of the Morley equation.

We will demonstrate that only the response to the higher modes $(n \geq 2)$ is markedly altered by the presence of an elastic support in comparison with a cylinder with a clamped or hinged support.

For these higher modes, the membrane force n_{xx} at the elastically supported base is directly related to the induced out-of-roundness ("ovalisation") of the cylinder, which is partly withstood at the base by the planar (circumferential and radial) elastic supports. As said before, the cross-section tends to warp at the base, which in turn is partly withstood by the axial elastic support and results in the normal stresses at the base. Elastic axial supports reduce warping less than rigid supports, n_{xx} remains smaller accordingly.

12.5.1 Derivation of Formulas

For a completely elastic supported edge, the following system of equations for the boundary conditions at the base $(x = 0)$ is obtained for the modes $n \geq 2$.

$$
\begin{bmatrix} k_x u_x \\ k_\theta u_\theta \\ k_z u_z \\ k_\varphi \varphi_x \end{bmatrix}^{x=0}
=
\begin{bmatrix} -n_{xx} \\ -n_{x\theta} \\ -v_x^* \\ -m_{xx} \end{bmatrix}^{x=0}
\tag{12.38}
$$

in which the spring stiffnesses k_x (axial), k_θ (circumferential), k_z (radial) and k_φ (rotational) are introduced.

   To solve this system, with the objective to obtain a formula for the stress at the base of the chimney, terms multiplied by $(a\beta)^{-2}$ are neglected in comparison to unity. We investigate some particular cases. As a reference, the results of Sect. 12.2 are recalled. The solution for a clamped base is thus obtained by equating each spring stiffness to infinity $(k_x = k_\theta = k_z = k_\varphi = \infty)$ and the solution for the "hinged-wall" edge $(u_x = u_\theta = u_z = 0,\ m_{xx} = 0)$ is thus obtained by equating each extensional spring stiffness to infinity and the rotational spring stiffness to zero $(k_x = k_\theta = k_z = \infty,\ k_\varphi = 0)$.

   We consider the following elastic support conditions:

1. Axial elastic support only,
2. Combination of axial and rotational elastic supports, and
3. Combination of circumferential and radial elastic supports.

   Identical to Sect. 12.2.1, we derive a design formula for the stress distribution at the base of the long chimney with elastic support by calculating the response to the wind load.

Axial Elastic Support

We assume that an axial elastic support k_x is present and that the wall of the cylinder is free to rotate. The displacements in the circular plane (θz-plane) are supposed to be fixed. We introduce the dimensionless parameter η_x, which is defined as

$$
\eta_x = \frac{k_x\, a}{E\, t}\, \frac{a\beta}{n\sqrt{n^2 - 1}}
\tag{12.39}
$$

   This parameter is mainly described by the geometrical properties of the cylinder and the ratio of the axial elastic support to the modulus of elasticity of the cylinder.

Now, the solution for this elastic supported edge $(k_x u_x = -n_{xx}, u_\theta = u_z = 0, m_{xx} = 0)$ reads

$$C_1^n = -\frac{v}{2}\frac{n^2-1}{(a\beta)^2}\frac{\eta_x}{\eta_x+1}\hat{u}_z^n; \quad C_2^n = 0;$$

$$C_5^n = -\hat{u}_z^n; \quad C_6^n = -\frac{\eta_x}{\eta_x+1}\hat{u}_z^n \tag{12.40}$$

By back substitution, we obtain the membrane force n_{xx} and the moment m_{xx}:

$$n_{xx}^{2 \le n \le 5}(0,\theta) \approx -\sum_{n=2}^{5} 2\sqrt{3(1-v^2)}\frac{a^2}{t}\frac{p_n}{n^2-1}\frac{\eta_x}{\eta_x+1}\cos n\theta \tag{12.41}$$

$$m_{xx}^{2 \le n \le 5}(0,\theta) = 0$$

Finally the axial stress distribution at the base becomes

$$\sigma_{xx}^{2 \le n \le 5}(0,\theta,z) = -\sum_{n=2}^{5} 2\sqrt{3(1-v^2)}\frac{a^2}{t^2}\frac{p_n}{n^2-1}\frac{\eta_x}{\eta_x+1}\cos n\theta \tag{12.42}$$

Now we introduce the normalised stress ratio λ_{xn}, which relates the stresses due to the higher modes to the same stresses that would occur for a rigid base

$$\lambda_{xn} = \frac{\sigma_{xx}^{0 \le n \le 5}(\eta_x) - \sigma_{xx}^{n=1}}{\sigma_{xx}^{0 \le n \le 5}(\eta_x = \infty) - \sigma_{xx}^{n=1}} = \frac{\sum\limits_{n=2}^{\infty}\sigma_{xx}^{2 \le n \le 5}(\eta_x)}{\sum\limits_{n=2}^{\infty}\sigma_{xx}^{2 \le n \le 5}} = \frac{\sum\limits_{n=2}^{5}\left(\frac{\alpha_n}{n^2-1}\frac{\eta_x}{\eta_x+1}\right)}{\sum\limits_{n=2}^{5}\left(\frac{\alpha_n}{n^2-1}\right)}$$

$$\tag{12.43}$$

Then, we can write the formula for the maximal tensile stress at the middle surface

$$\sigma_{xx,t}^{0 \le n \le 5}(z=0) = 0.224\frac{l^2}{at}p_w\left[1+6.39\sqrt{1-v^2}\lambda_{xn}\left(\frac{a}{l}\right)^2\frac{a}{t}\right] \tag{12.44}$$

Axial and Rotational Elastic Support

We assume that, next to the axial elastic support k_x, a rotational elastic support k_φ is present. The displacements in the circular plane (θz-plane) are supposed to be fixed, which is equal to the previous case.

We introduce the dimensionless parameters η_x and η_φ as

$$\eta_x = \frac{k_x\,a}{E\,t}\frac{a\beta}{n\sqrt{n^2-1}}; \quad \eta_\varphi = \frac{k_\varphi\,a}{E\,t}2\beta^2 \tag{12.45}$$

Then, the solution for this elastic supported edge ($k_x u_x = -n_{xx}, u_\theta = u_z = 0$, $k_\varphi \varphi_x = -m_{xx}$) reads

$$C_1^n = -\frac{v\,n^2-1}{2\,(a\beta)^2}\frac{\eta_x}{\eta_x+1}\hat{u}_z^n$$

$$C_2^n = -\frac{\eta_\varphi}{\eta_\varphi+1}\left[\frac{v\,n^2-1}{2\,(a\beta)^2}\frac{\eta_x}{\eta_x+1}+\frac{1}{2}\frac{n\sqrt{n^2-1}}{(a\beta)^2}\left(1-\frac{\eta_x}{\eta_x+1}\right)\right]\hat{u}_z^n \tag{12.46}$$

$$C_5^n = -\hat{u}_z^n$$

$$C_6^n = -\frac{\eta_x}{\eta_x+1}\hat{u}_z^n$$

By back substitution, we obtain the membrane force n_{xx} and the moment m_{xx}:

$$n_{xx}^{2\le n\le 5}(0,\theta)\approx-\sum_{n=2}^{5}2\sqrt{3(1-v^2)}\frac{a^2}{t}\frac{p_n}{n^2-1}\frac{\eta_x}{\eta_x+1}\cos n\theta$$

$$m_{xx}^{2\le n\le 5}(0,\theta)\approx-\sum_{n=2}^{5}a^2\frac{p_n}{n^2-1}\frac{\eta_\varphi}{\eta_\varphi+1}\left[v\frac{\eta_x}{\eta_x+1}+\frac{n^2}{n\sqrt{n^2-1}}\left(1-\frac{\eta_x}{\eta_x+1}\right)\right]\cos n\theta \tag{12.47}$$

If the parameter η_x is large and thus the factor $\eta_x/(\eta_x+1)$ close to unity, the stress distribution at the base can be obtained by

$$\sigma_{xx}^{2\le n\le 5}(0,\theta,z)=\frac{n_{xx}}{t}+\frac{2z}{t}\frac{6m_{xx}}{t^2}$$

$$=-\sum_{n=2}^{5}2\sqrt{3(1-v^2)}\frac{a^2}{t^2}\frac{p_n}{n^2-1}\frac{\eta_x}{\eta_x+1}\left(1+\frac{2z}{t}\sqrt{3}\frac{v}{\sqrt{1-v^2}}\frac{\eta_\varphi}{\eta_\varphi+1}\right)\cos n\theta \tag{12.48}$$

If the parameter η_x is not large, the parameter η_φ is probably small in the practical cases and hence the moment m_{xx} is almost zero.

The stress at the middle surface is for all cases described by

$$\sigma_{xx}^{2\le n\le 5}(0,\theta,0)=-\sum_{n=2}^{5}2\sqrt{3(1-v^2)}\frac{a^2}{t^2}\frac{p_n}{n^2-1}\frac{\eta_x}{\eta_x+1}\cos n\theta \tag{12.49}$$

Based on the closed-form solution and practical considerations, we assess that the additional influence of the rotational support will be limited, as the rotational spring stiffness k_φ will decrease rapidly with decreasing axial spring stiffness k_x.

Circumferential and Radial Elastic Support

We assume that both a circumferential elastic support k_θ and a radial elastic support k_z are present. The displacement in axial direction is supposed to be fixed, while the wall of the cylinder is free to rotate.

The solution for this elastic supported edge $(u_x = 0, k_\theta u_\theta = -n^*_{x\theta}, k_z u_z = -v^*_x, m_{xx} = 0)$ reads

$$C^n_1 \approx C^n_2 \approx O\left((a\beta)^{-2}\right); \quad C^n_5 \approx C^n_6 \approx -\frac{\eta_{\theta z}}{\eta_{\theta z} + 1}\hat{u}^n_z \qquad (12.50)$$

We now introduce the dimensionless parameter $\eta_{\theta z}$:

$$\eta_{\theta z} = 2\frac{n^2 k_z + k_\theta}{E}\frac{a}{t}\frac{1}{n}\left(\frac{a\beta}{\sqrt{n^2 - 1}}\right)^3 \qquad (12.51)$$

Again, this parameter is mainly described by the geometrical properties of the cylinder and the ratio of the combined elastic support to the modulus of elasticity of the cylinder. The approximate solution above is accurate if the parameter $\eta_{\theta z}$ is not small, since then the moment m_{xx} is almost zero and does not exert influence on the stress distribution at the base. By back substitution, we obtain the membrane force n_{xx} and the moment m_{xx}:

$$n^{2 \leq n \leq 5}_{xx}(0, \theta) \approx -\sum_{n=2}^{5} 2\sqrt{3(1 - v^2)}\frac{a^2}{t}\frac{p_{zn}}{n^2 - 1}\frac{\eta_{\theta z}}{\eta_{\theta z} + 1}\cos n\theta;$$

$$m^{2 \leq n \leq 5}_{xx}(0, \theta) \approx 0 \qquad (12.52)$$

Finally, the axial stress distribution at the base becomes

$$\sigma^{2 \leq n \leq 5}_{xx}(0, \theta, z) = -\sum_{n=2}^{5} 2\sqrt{3(1 - v^2)}\frac{a^2}{t^2}\frac{p_{zn}}{n^2 - 1}\frac{\eta_{\theta z}}{\eta_{\theta z} + 1}\cos n\theta \qquad (12.53)$$

Introducing the normalised parameter $\lambda_{\theta zn}$,

$$\lambda_{\theta zn} = \frac{\sigma^{0 \leq n \leq 5}_{xx}(\eta_{\theta z}) - \sigma^{n=1}_{xx}}{\sigma^{0 \leq n \leq 5}_{xx}(\eta_{\theta z} = \infty) - \sigma^{n=1}_{xx}} = \frac{\sum_{n=2}^{\infty} \sigma^{2 \leq n \leq 5}_{xx}(\eta_{\theta z})}{\sum_{n=2}^{\infty} \sigma^{2 \leq n \leq 5}_{xx}} = \frac{\sum_{n=2}^{5}\left(\frac{\alpha_n}{n^2-1}\frac{\eta_{\theta z}}{\eta_{\theta z}+1}\right)}{\sum_{n=2}^{5}\left(\frac{\alpha_n}{n^2-1}\right)}$$

$$(12.54)$$

we obtain the formula for the maximal tensile stress at the middle surface:

$$\sigma^{0 \leq n \leq 5}_{xx,t}(z = 0) = 0.224\frac{l^2}{at}p_w\left[1 + 6.39\sqrt{1 - v^2}\,\lambda_{\theta zn}\left(\frac{a}{l}\right)^2\frac{a}{t}\right] \qquad (12.55)$$

12.5.2 Applicability Range of Formulas

Similar to Sect. 12.2, Eqs. (12.49) and (12.53) predict the tensile axial stress at the base and the windward side of a long chimney subject to wind load. However, the influence of elastic supports is incorporated into the formulas.

We have investigated the range of application of these formulas by comparison with results obtained by the developed program, which applies for short and long cylindrical shells and allows accurate modelling of elastic supports. As this program is based on the closed-form solution, an identical result is obtained for chimneys much longer than the influence length. For chimneys shorter than the influence length, the program is more accurate since the formulas do not include the effect of the edge disturbance that originates at the free edge.

We have performed calculations for both axial elastic supported cylinders and cylinders with both a circumferential elastic support and a radial elastic support. The approach and results are discussed below.

Axial Elastic Support

For the axial elastic support (described by η_x), calculations have been made for a constant radius-to-thickness-ratio equal to 100 and length-to-radius-ratios of 5, 10, 20 and 40. This range has been chosen based on the dominating long influence length ($n = 2$) for the radius-to-thickness-ratio equal to 100, which is approximately equal to 20 times the radius. The parameter η_x has been varied from practically infinity to zero in combination with a "hinged wall", i.e., k_φ equal to zero and the total-stress-to-beam-stress-ratio has been investigated. The total-to-beam-stress-ratio for the elastically supported condition is normalised to the "hinged-wall" solution. Obviously, if the parameter η_x is large, this "hinged-wall" solution is obtained and, if the parameter η_x approaches zero, the total-stress-to-beam-stress-ratio is equal to unity.

The agreement between the theoretical factor and the factors calculated by the developed program is very good and excellent for the higher length-to-radius ratios. Based on these results, it has been demonstrated that the closed-form solution is thus applicable for any value of the parameter η_x, which expresses the stiffness of the axial elastic support, and if the length of the chimney is longer than half of the influence length for mode number $n = 2$, which coincides with the range of application of Eqs. (12.20) and (12.23) for a fixed base. In other words, Eq. (12.49) that additionally accounts for the presence of an axial elastic support has the same range of application as the formula without this additional term.

Circumferential and Radial Elastic Support

For the circumferential and radial axial elastic support (described by $\eta_{\theta z}$), it has been assumed that the circumferential spring stiffness k_θ is equal to the radial spring stiffness k_z and, as such, a planar elastic support is provided. Calculations have been made for a radius-to-thickness-ratio of 100 in combination with length-to-radius-ratios of 2.5, 5, 10 and 20, respectively. This range has been chosen based on the influence length for the radius-to-thickness-ratio equal to 100, which is approximately equal to 20 times the radius. Longer chimneys are not considered since, from the previous results for the axial elastic support described with the parameter η_x, it can be concluded that the formula is valid for chimneys longer than the influence length. The parameter $\eta_{\theta z}$ has been varied from practically infinity to zero in combination with a "hinged wall", i.e. k_θ and k_z equal to zero and the total-stress-to-beam-stress-ratio has been investigated. The total-to-beam-stress-ratio for the elastically supported condition is normalised to the "hinged-wall" solution. Obviously, if the parameter $\eta_{\theta z}$ is large, this "hinged-wall" solution is obtained and, if the parameter $\eta_{\theta z}$ approaches zero, the total-stress-to-beam-stress-ratio is equal to unity.

In comparison with the influence of an axial elastic support, it has been observed that the agreement between the theoretical factor and the factors calculated by the developed program is even better than for the variation of the parameter η_x. The agreement is even quite good for a length-to-radius-ratio of 2.5, which is much less than half of the influence length (here $l_{i,2}^{n=2} \big/ 2a \approx 10$ for $a/t = 100$).

Since the formula for the stress at the base of a chimney is not accurate for a length smaller than the half influence length, it is remarkable that the closed form solution for the influence of an elastic supported edge is even more accurate. Additionally, smaller values than 2.5 for the length over the radius are not practical from an engineering point of view. However, the range of application for the total stress at the elastic supported base of a chimney loaded by the wind load is governed by the limitations of the formula for the clamped or hinged supported base.

In Fig. 12.5, the normalized curves for the theoretical factor obtained with Eqs. (12.49) and (12.53) are shown. On the horizontal axis, the modified parameters $\eta_{x,\mathrm{mod}}$ and $\eta_{\theta z,\mathrm{mod}}$ have been adopted. The parameter $\eta_{x,\mathrm{mod}}$ is in fact a reduction of the parameter η_x according to

$$\eta_{x,\mathrm{mod}} = \eta_x \frac{n\sqrt{n^2-1}}{\sqrt[4]{3(1-v^2)}} = \frac{k_x}{E} \frac{a}{t} \sqrt{\frac{a}{t}} \qquad (12.56)$$

This modified parameter is thus independent of the mode number n, while the dependency on the radius-to-thickness-ratio and the elastic properties of the chimney is preserved.

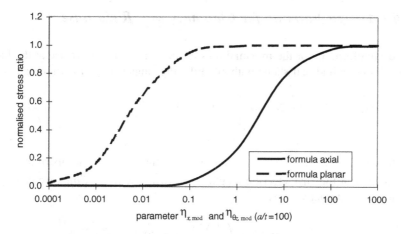

Fig. 12.5   Theoretical factor for axial elastic support and planar elastic support

The parameter $\eta_{\theta z,mod}$ is a modification of the parameter $\eta_{\theta z}$ in Eq. (12.51) according to

$$\eta_{\theta z,mod} = \eta_{\theta z} \frac{1}{2(n^2 + 1)} \, \eta_n \frac{n^2 - 1}{\sqrt[4]{3(1 - v^2)}} = \frac{k_{\theta z}}{E} \frac{a}{t} \sqrt{\frac{a}{t}} \qquad (12.57)$$

in which $k_{\theta z} = k_\theta = k_z$. The parameter $\eta_{\theta z}$ is thus modified corresponding to the modification of the parameter η_x to show the relative influence of the parameters.

Based on Fig. 12.5, it is concluded that in case of an elastic support to a long circular cylinder, only the axial spring stiffness has to be taken into account as the influence of the planar spring stiffness is only markedly observed for very low stiffness ratios in comparison to the axial stiffness ratio.

12.6   Summary of Chimney Design Formulas

In this chapter, we have presented the design formulas that are based on the Morley equation and an equation on the basis of the semi-membrane concept. We also indicated the respective ranges of application. The main objective of the chapter was the design of chimneys, but the derived formulas are meaningful for any slender circular cylindrical shell structure.

12.6.1 Design Formula for Chimneys with Rigid Base

The design formula for the maximal tensile stress at the outer surface at the fixed base of long cylindrical shells without stiffening rings subject to wind load is

$$\sigma_{xx,t}^{0 \leq n \leq 5}(z = t/2) = 0.224 \frac{l^2}{at} p_w \left[1 + 6.39\sqrt{1 - v^2} \left(\frac{a}{l} \right)^2 \frac{a}{t} \right] \cdot \left(1 + \sqrt{3} \frac{v}{\sqrt{1 - v^2}} \right)$$

$$(12.58)$$

The coefficients 0.224 and 6.39 hold true for the chosen wind distribution. By introduction of the characteristic lengths $l_{c1} = \sqrt{at}$ and $l_{c2} = \sqrt[4]{atl^2}$, we can write this formula alternatively as

$$\sigma_{xx,t}^{0 \leq n \leq 5}(z = t/2) = 0.224 p_w \left(\frac{l}{l_{c1}} \right)^2 \left[1 + 6.39\sqrt{1 - v^2} \left(\frac{a}{l_{c2}} \right)^4 \right]$$
$$\cdot \left(1 + \sqrt{3} \frac{v}{\sqrt{1 - v^2}} \right) \qquad (12.59)$$

The term within the round brackets describes the effect of a full rotational constraint and is omitted in case the shell wall is free to rotate at the base.

The formula is applicable to cylinders longer than half of the long influence length. Alternatively, one can state that the formula is applicable to cylinders with a characteristic length l_{c2} longer than the radius a.

12.6.2 Design Formula for Chimney with Stiffening Rings

The design formula for the maximal tensile stress reads

$$\sigma_{xx,t}^{0 \leq n \leq 5}(z = t/2) = 0.224 \frac{l^2}{at} p_w \left[1 + 6.39\sqrt{1 - v^2}\sqrt{\lambda_r} \left(\frac{a}{l} \right)^2 \frac{a}{t} \right]$$
$$\cdot \left(1 + \sqrt{3} \frac{v}{\sqrt{1 - v^2}} \right) \qquad (12.60)$$

or alternatively

$$\sigma_{xx,t}^{0 \leq n \leq 5}(z = t/2) = 0.224 p_w \left(\frac{l}{l_{c1}} \right)^2 \left[1 + 6.39\sqrt{1 - v^2}\sqrt{\lambda_r} \left(\frac{a}{l_{c2}} \right)^4 \right]$$
$$\cdot \left(1 + \sqrt{3} \frac{v}{\sqrt{1 - v^2}} \right) \qquad (12.61)$$

in which the stiffness ratio λ_r represents the ratio of the bending stiffness of the circular cylindrical shell only to the modified bending stiffness of the shell (including the 'smeared' contribution of the ring stiffness).

The design formula is applicable for closely spaced stiffening rings, i.e. with a spacing shorter than half of the long influence length.

As a rule, ring stiffeners are connected eccentrically. An effective width of the shell must be taken into account for calculating the flexural stiffness of the stiffeners.

12.6.3 Design Formula for Chimneys with Elastic Supports

The design formula for the maximal tensile stress at the middle surface at the fixed base of long chimneys with axial elastic support subject to wind load is

$$\sigma_{xx,t}^{0 \le n \le 5}(z=0) = 0.224\frac{l^2}{at}p_w\left[1 + 6.39\sqrt{1-v^2}\lambda_{xn}\left(\frac{a}{l}\right)^2\frac{a}{t}\right] \qquad (12.62)$$

or alternatively

$$\sigma_{xx,t}^{0 \le n \le 5}(z=0) = 0.224p_w\left(\frac{l}{l_{c1}}\right)^2\left[1 + 6.39\sqrt{1-v^2}\lambda_{xn}\left(\frac{a}{l_{c2}}\right)^4\right] \qquad (12.63)$$

in which the normalised stress ratio λ_{xn} is introduced, defined in Eq. (12.43).

Similarly, the design formula for the maximal tensile stress at the middle surface in case of combined circumferential and radial elastic support is

$$\sigma_{xx,t}^{0 \le n \le 5}(z=0) = 0.224\frac{l^2}{at}p_w\left[1 + 6.39\sqrt{1-v^2}\lambda_{\theta zn}\left(\frac{a}{l}\right)^2\frac{a}{t}\right] \qquad (12.64)$$

or alternatively

$$\sigma_{xx,t}^{0 \le n \le 5}(z=0) = 0.224p_w\left(\frac{l}{l_{c1}}\right)^2\left[1 + 6.39\sqrt{1-v^2}\lambda_{\theta zn}\left(\frac{a}{l_{c2}}\right)^4\right] \qquad (12.65)$$

in which the normalised stress ratio $\lambda_{\theta zn}$ is introduced, defined in Eq. (12.54). The formulas are applicable to cylinders for which the characteristic length l_{c2} is larger or equal to its radius a. In case of an elastic support to a long circular cylinder, we must take into account only the axial spring stiffness and the formula is applicable for any value of the parameter η_x and cylinders longer than half of the long influence length.

Chapter 13
Storage Tanks

To demonstrate the capability of the developed program CShell, we have performed a numerical study of tanks under the main load-deformation conditions. This chapter focuses on the shell of the tank while considering the various connections of the shell to the top and bottom, i.e., the influence of type of roof and floor on the behavior of the tank wall.

13.1 Problem Statement

Circular cylindrical tanks are used for storing liquids, gases, solids and mixtures thereof. Tanks for storing solids are more usually referred to as silos. Tanks forming a closed container designed to hold gases or liquids at a pressure substantially different from the ambient pressure are referred to as pressure vessels. Silos and pressure vessels are not considered in this chapter.

Liquid storage tanks for storing water, oil, fuel, chemicals and other fluids are usually vertical in shape. A typical large, thin-walled liquid storage tank is obviously much shorter than the long chimney such that the diameter is of the same order in comparison with the length as opposed to the chimney. The geometry of such stocky cylinders is typically such that the diameter is at least equal to the length or that the length can even be much smaller than the radius, viz. a ratio between radius and length between 0.5 and 3.

For such short lengths between the circular boundaries, the short influence length has a more marked contribution to the load-deformation behavior of the cylinder and the long influence length is much longer than the height of the shell. This feature prevents us from readily obtaining a closed-form solution to non-axisymmetric loads similar to those obtained for the long chimneys in Chap. 12.

Concrete tanks typically might have a relatively large ratio between radius and thickness of about 30–80, but especially large steel storage tanks are thin-walled

J. Blaauwendraad and J. H. Hoefakker, *Structural Shell Analysis*,
Solid Mechanics and Its Applications 200, DOI: 10.1007/978-94-007-6701-0_13,
© Springer Science+Business Media Dordrecht 2014

such that the ratio between radius and thickness might even be between 500 and 2,000.

In this chapter, we intend primarily to demonstrate the capability of the program developed to model the shell of large vertical liquid storage tanks. Additionally, we provide tentative insight into the static response of such tank shells to the relevant load and/or deformation conditions. This is obtained by several calculations with the program CShell and comparison with the insight as obtained for the behavior of the long cylinder.

This chapter further focuses primarily on large, single wall, concrete or steel, vertical tanks, which are either closed or open at the top, for the storage of liquids at low or ambient temperatures and with a design pressure near ambient pressure. The design of such tanks can be divided into three major areas: (1) the shell, (2) the bottom, and (3) the roof. The bottom and roof layout of the tank typically vary with the operational conditions, preferences and safety requirements. In any case, these provide a rigid support, no support, or an elastic (intermediate) support to the tank shell. In view of the capability of the developed program, the next sections focus on the shell of the tank while considering the various connections of the shell to the top and bottom.

13.2 Load-Deformation Conditions and Analysed Cases

We can identify three main load-deformation conditions for the overall response of the tank wall:

- Content load (especially when being filled to maximum capacity),
- Wind load (especially for the open top tank and external floating roof tank),
- Settlement induced load and/or deformation.

The response to the content load has been described in Chap. 5 and the possible effects of non-uniform settlement of the foundation on the deformation of the tank wall have been treated in Sect. 4.8.

As described in Sect. 12.1, the shape of the circumferential distribution of the wind load is such that only a part of the circumference, the so-called stagnation zone, is under circumferential compression, while the remainder is under suction. It is noted for short cylinders that we should consider different values depending on the Reynolds number corresponding to the geometry of the cylinder. Here we adopt for convenience the wind load distribution as described by Eq. (12.1).

Due to the wind load, the cross-section of the tank tends to distort into an oval shape. At the shell-to-bottom junction, the full circular restraint of the tank wall in combination with an axial restraint in case of anchorage, induces axial bending stresses along the long influence length that result from the withstood out-of-roundness as similarly observed for the chimney under wind load.

In case of closed and fixed roof tanks, the roof provides an adequate restraint to maintain the roundness of the tank. The wind load is then mainly carried by axial tensile stresses at the windward side and compressive stresses at the leeward side, i.e., mainly by beam action of the shell. For open top and external floating roof tanks, circumferential primary wind girders are normally externally provided at or near the tank top to maintain the roundness and stability of the tank under wind load (especially while emptying the tank). Especially for tall tanks, secondary wind girders at intervals in the height of the tank might be required to prevent local buckling.

Based on the above main load-deformation conditions for the tank, the following relevant cases have been identified for the shell of the tank:

- Content load of a fully filled tank,
- Wind load on the tank with various top restraints, and
- Circumferential settlement of the foundation with various top restraints.

For the hydrostatic load and the wind load, both steel tanks and concrete tanks have been analyzed, while only steel shells have been considered for the settlement analyses.

13.3   Stresses Due to Content

13.3.1   Concrete Tank

For the hydrostatic load on a tank wall with uniform thickness, we can obtain the solution by elementary analysis as performed in Sect. 5.3 in which the response of a concrete tank is treated.

13.3.2   Steel Tank

For practical reasons, steel tanks are built up from fairly small rectangular pieces of carbon steel plate, which are curved in a cylindrical shape and joined by butt-welding. The shell is thus built up in rings (also referred to as courses) and typically the thickness of the plates varies with the internal pressure, i.e., thicker plates are applied in the lower courses and thinner plates in the upper courses.

A steel tank (material properties taken as $E = 210 \times 10^6 \, \text{kN/m}^2$ and $v = 0.3$) with radius $a = 10$ m, height $l = 20$ m and varying wall thickness is modeled as completely filled with water (density taken as $\gamma_w = 10 \, \text{kN/m}^3$). The bottom edge $(x = 0)$ is fully fixed and the top $(x = l)$ is free. The thickness of the shell courses has been applied as follows from the bottom to the top (h indicates height of the respective courses or courses with the same thickness):

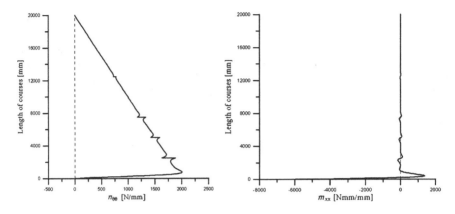

Fig. 13.1 Steel tank with content load, stress resultants $n_{\theta\theta}$ and m_{xx} along the height

$$
\begin{aligned}
h &= 2.5\,\text{m} &\times& \quad t = 11\,\text{mm} \\
h &= 2.5\,\text{m} &\times& \quad t = 9.5\,\text{mm} \\
h &= 2.5\,\text{m} &\times& \quad t = 8.5\,\text{mm} \\
h &= 5.0\,\text{m} &\times& \quad t = 7.5\,\text{mm} \\
h &= 7.5\,\text{m} &\times& \quad t = 7.0\,\text{mm}
\end{aligned}
\tag{13.1}
$$

The relevant displacements and stresses are shown in Figs. 13.1 and 13.2. The membrane force $n_{\theta\theta}$ in Fig. 13.1 varies rather linearly with the content level up to the region near the bottom where the radial displacement is a fully restraint. The small disturbances coincide with the transitions in course thickness. The axial stress associated with the moment m_{xx} in Fig. 13.1 is quite considerable, but in fact a fully fixed tank is modeled. If the tank is not fully fixed, an elastic rotational support is present that allows some rotation of the shell-to-floor junction, which effectively

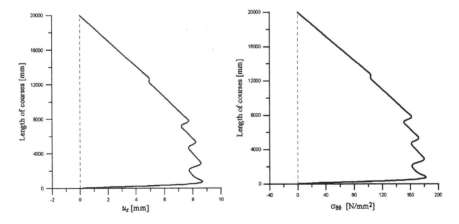

Fig. 13.2 Steel tank with content load, displacement u_z and hoop stress $\sigma_{\theta\theta}$ along the height

reduces the bending stresses at the bottom. The distribution of the hoop stress $\sigma_{\theta\theta}$ and the displacement u_z in Fig. 13.2 are identical, as expected. The shape of the hoop stress diagram is reduced by the increased thickness of the courses toward the bottom of the tank. The smooth changes are clarified by the stiffening effect of the thicker plate to the thinner plate above, which can be considered as a partial restraint at the bottom edge of the thinner plate.

13.4 Stresses Due to Wind Load

13.4.1 Concrete Tank

In this subsection, we present the response of two different concrete storage tanks under the wind load of Eq. (12.1). The two cases are:

1. A storage tank, which is clamped at the base, with a free edge at the top; and
2. The same storage tank, but with a fully rigid roof at the top.

The rigid roof is modeled as a ring with a very high modulus of elasticity so that the ring is non-deformable by in-plane actions and thus provides a circular restraint to the top. At the bottom, we consider a full axial and rotational restraint to the shell wall (i.e., clamped condition). The geometrical properties of both concrete shells are the same ($l = 30$ m, $a = 25$ m, $t = 0.3$ m) and for the material properties we use $E = 35 \times 10^6 \, \mathrm{kN/m^2}$ and $v = 0.2$.

From Fig. 13.3 we observe that, in view of the magnification factors, the deformation is drastically reduced by the rigid roof. Figure 13.4 reveals that the

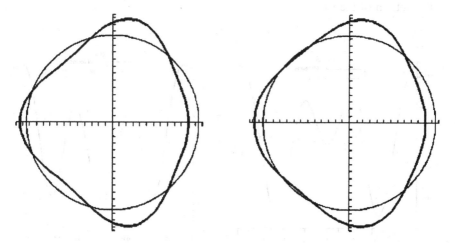

Fig. 13.3 Cross sectional deformation of the clamped tank with a free edge (*left*) at the top ($\times$ 3,000) and of the clamped tank with a rigid roof (*right*) at half the length of the cylinder ($\times$ 18,000)

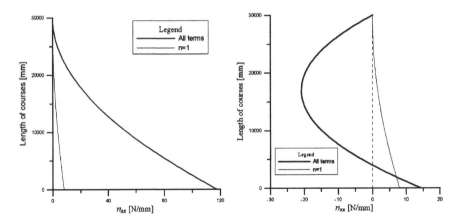

Fig. 13.4 Clamped tank with a free edge (*left*) and with a rigid roof (*right*), n_{xx} at $\theta = 0$

axial membrane forces n_{xx} for $n = 1$ and $n > 1$ are distributed like a clamped-free beam. In case of a rigid roof, this again applies for $n = 1$, but not anymore for $n > 1$. Then the stress is distributed like a clamped-hinged beam. At the base of the shell with the rigid roof, the membrane force n_{xx} is much smaller than at the base with the free edge. The distribution of this membrane force along the base circle is shown in Fig. 13.5. Finally, the membrane force n_{xx} at the base of the shell with the free edge is about 120 N/mm under the applied wind load. This value is even less than the axial stress at the base under the dead weight of the concrete shell only. Adopting a typical density of 2400 kg/m^3 for concrete, the dead weight membrane force at the base becomes $\rho g l t \approx 210$ N/mm. In this particular case, it can thus be concluded that the dead weight virtually provides a full axial restraint at the bottom of the shell.

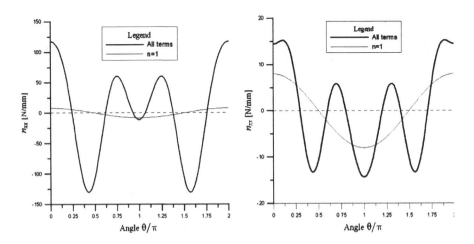

Fig. 13.5 Clamped tank with a free edge (*left*) and with a rigid roof (*right*), n_{xx} at $x = 0$

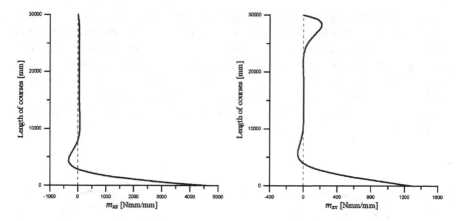

Fig. 13.6 Clamped tank with a free edge (*left*) and with a rigid roof (*right*), m_{xx} at $\theta = 0$

For reasons of comparison we analyzed the same tank on the basis of the Donnell equation. The magnitude of the stress resultants and cross-sectional deformation is much smaller in the Morley theory. The membrane force n_{xx} is reduced by about 20 % at the base of the tank with the free edge and, in case of a rigid roof, the maximum along the tank height is reduced by about 33 %. Moreover, the moment m_{xx} at the base is reduced by 50 %. For this particular case, a large reduction in the outer fiber stress at the base is thus observed. The quantities and the shape of the diagrams are properly described by the Donnell equation, but to predict the magnitude of these quantities, we should consider the Morley equation.

The bending moment m_{xx} is shown in Fig. 13.6. This moment only occurs in the base zone and top zone due to short edge disturbances. Over a large part of the shell the moment is practically equal to zero. Due to the deformed shape in the horizontal plane, see Fig. 13.3, the bending moment $m_{\theta\theta}$ will be nonzero and will be substantial for a free edge.

13.4.2 Steel Tank

In this subsection, the response of two different steel storage tanks (material properties taken as $E = 210 \times 10^6 \, \mathrm{kN/m^2}$ and $v = 0.3$) under the wind load of Eq. (12.1) is presented. The two cases are:

1. A storage tank, which is clamped at the base, with a free edge at the top; and.
2. The same storage tank, but with a (steel) wind girder at the top.

In line with the observations of the previous subsection, the connection at the base is modeled as a full axial and rotational restraint to the shell wall. These cases are considered to show the impact of the wind girder on the stress distribution and the deformation of the tank.

A typical geometry for a steel storage tank with a wind girder is $a/t = 1000$, $l/a = 1$ and $\lambda_g = 20$ where the ratio λ_g represents the bending rigidity of the wind girder itself to the tank wall. Hence, a tank with $l = a = 10$ m is considered that is built up from various courses with varying plate thickness as exemplified in Sect. 13.3.2 while maintaining roughly the typical ratio of $a/t = 1,000$. The thickness of the shell courses has been applied as follows from the bottom to the top (h indicates height of the respective courses or courses with the same thickness):

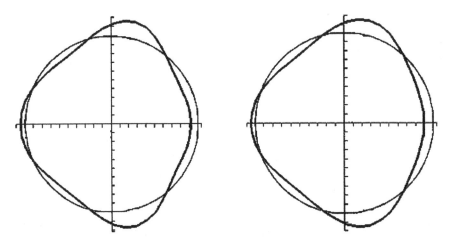

Fig. 13.7 Cross-sectional deformation at the top of a steel tank with a free edge (*left*) ($\times$ 2,000) and with a wind girder (*right*) ($\times$ 2,000)

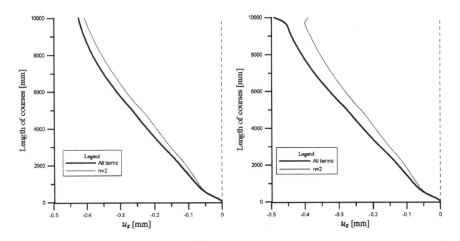

Fig. 13.8 Steel tank with a free edge (*left*) and with a wind girder (*right*), u_z at $\theta = 0$

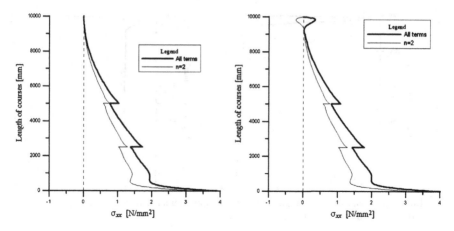

Fig. 13.9 Steel tank with a free edge (*left*) and with a wind girder (*right*), σ_{xx} at $\theta = 0$ and at outer face of the cylinder

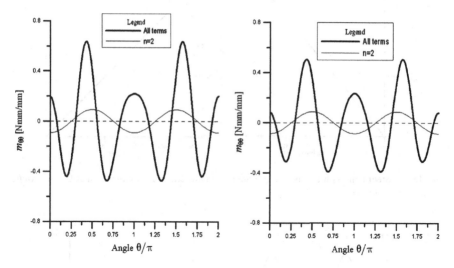

Fig. 13.10 Steel tank with a free edge (*left*) and with a wind girder (*right*), $m_{\theta\theta}$ at the *top*

$$h = 2.5\,\text{m} \quad \times \quad t = 12.5\,\text{mm}$$
$$h = 2.5\,\text{m} \quad \times \quad t = 10.0\,\text{mm} \tag{13.2}$$
$$h = 5.0\,\text{m} \quad \times \quad t = 7.5\,\text{mm}$$

The corresponding tank wall bending stiffness (viz. shell bending rigidity times the tank height) is thus equal to.

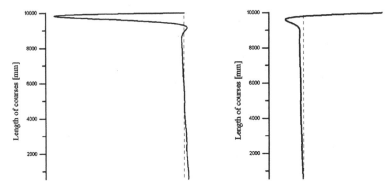

Fig. 13.11 Steel tank with a wind girder under a circumferential settlement, σ_{xx} at $\theta = 0$ and $\sigma_{\theta\theta}$ at $\theta = 0$ (*right*) and both at outer face of the cylinder

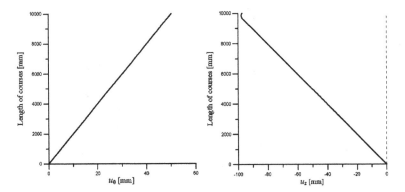

Fig. 13.12 Steel tank with a wind girder under a circumferential settlement, u_θ at $\theta = 45°$ (*left*) and u_z at $\theta = 0$ (*right*)

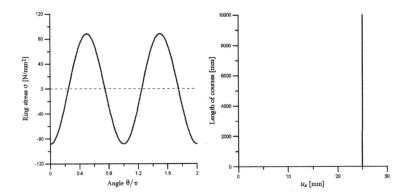

Fig. 13.13 Steel tank with a wind girder under a circumferential settlement, circumferential stress σ_{ring} (*left*) and u_x at $\theta = 0$ (*right*)

$$D_b l = \frac{E}{12(1 - v^2)} \sum h t^3 \qquad (13.3)$$

For the present purpose, the wind girder is conveniently modeled as an eccentric annular plate with width $h_g = 250$ mm and thickness $t_g = 12.5$ mm resulting in a circumferential bending rigidity of the wind girder $EI_g = E h_g^3 t_g / 12$ with respect to its centre of gravity.

From Figs. 13.7 up to 13.9 and including Fig. 13.10, we observe that the influence of the modeled ring is confined to a limited length from the top, viz. only affects the shell behavior within the short influence length and does not markedly influence the overall behavior. Note in Fig. 13.7 that the radial displacement u_z does not differ much between shells with or without a wind girder. The positive effect of a wind girder in carrying the wind load is very small (Figs. 13.11–13.13).

13.5 Settlement Induced Stresses

In this section, we present the response of the steel tank with the wind girder of the previous subsection (case 2) under a non-planar settlement. For the present purpose, we conveniently introduce a full circumferential settlement with mode number $n = 2$ as done in Sect. 4.8, which is described by

$$u_x(\theta) = -u_{s,max} \cos 2\theta \qquad (13.4)$$

in which $u_{s,max}$ is the maximum settlement along the circumference. Here the value $u_{s,max} = 25$ mm is chosen as typical value. The geometry and material properties are taken identical to those of the previous subsection, but at the bottom we model the tank as freely supported in axial and rotational direction. In other words, at the bottom of the tank a prescribed axial displacement without rotational constraint is modeled.

In Sect. 4.8, an assessment based on in-extensional behavior of a cylinder with height equal to its radius a, subject to a general sagging of the tank ($n = 2$), is performed. From Eq. (4.53), we find that the radial displacement u_z at the top of such a cylinder is equal to four times the axial displacement u_x of the settlement and two times the circumferential displacement u_θ along the shell height.

From the Figs. 13.11–13.13, we observe that the circumferential settlement indeed mainly induces an in-extensional deformation and corresponding stresses. Furthermore, we observe that the influence of the modeled ring is confined to a limited length from the top, viz. only affects the shell behavior within the short influence length and does not markedly influence the overall behavior. Hence, the ring at the top only influences the hoop stress and the axial stress and hardly affects the deformation. The ring just follows the in-extensional behavior and the bending moment is fully deformation-imposed.

Part IV
Cones and Spheres

Chapter 14
Membrane Behaviour of Shells of Revolution Under Axisymmetric Loading

For roof shells with a rectangular plan it is convenient to apply a general co-ordinate system (x, y, z). This description is useful for a *shallow shell*: in other words for a shell with a span much larger than the elevation. For *deep shells* this description is not accurate. For manual calculation of the special group of non-shallow shells of revolution, we would do well to apply another description. This chapter deals with the membrane behaviour of these shells under axisymmetric loading.

14.1 Description of the Surface

Figure 14.1 shows a shell of revolution with a vertical axis of revolution. A point P on the surface of the shell lies on a *meridian* and a *parallel* or *latitude circle*. The meridian is the intersection line of the shell and the vertical plane through the axis of revolution and Point P. The latitude circle is the intersection line of the shell and the horizontal plane through P. The normal to the shell surface in point P intersects the axis of revolution at point Q. In Fig. 14.1a, we have drawn a set of axes x, y, z in point P. The x-axis is along the meridian, which is in fact the longitudinal direction. The y-axis is along the latitude circle and the z-axis along the normal. It is convenient to replace the set of axes by an alternative set of polar coordinates as shown in Fig. 14.1b. In the vertical plane, the angle ϕ between the normal and axis of rotation takes the place of x, and the angle θ in the horizontal plane replaces y. The z-axis remains unchanged in the normal direction.

Shells of revolution have principal curvatures in the direction of the meridians and latitude circles. The origins of the two radii of curvature are on the normal, but need not be at the same place. The principal radius r_1 corresponds to the curvature in the vertical plane, see Fig. 14.1c. The principal radius of curvature r_2 in the other direction is shown in Fig. 14.1d. This radius starts at point Q on the axis of revolution; the origin of radius r_1 is not restricted to point Q. That is only the case for spheres. Positive directions of the displacements of a point P on a meridian are

J. Blaauwendraad and J. H. Hoefakker, *Structural Shell Analysis*,
Solid Mechanics and Its Applications 200, DOI: 10.1007/978-94-007-6701-0_14,
© Springer Science+Business Media Dordrecht 2014

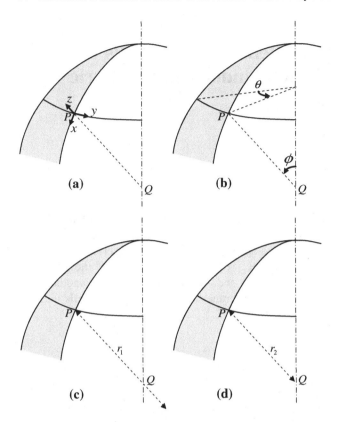

Fig. 14.1 Coordinates in two systems, (a) and (b). Principal radii, (c) and (d)

shown in Fig. 14.2. Left shows the displacements as defined in the direction of the coordinates ϕ and z; right shows the displacement in the direction of the horizontal radius r. Because of axisymmetry, the load p_θ, the displacement u_θ and all derivatives with respect to θ are zero. Since the edges of a shell of revolution are often situated in the θr-plane, it is useful to introduce the alternative displacement u_r as shown in Fig. 14.2. For membrane states, the following vectors are relevant:

Fig. 14.2 Displacements in the vertical plane

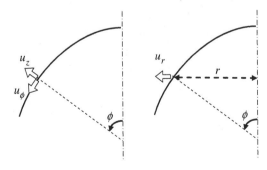

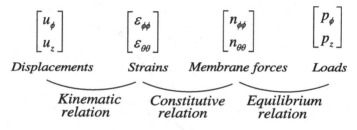

$$\begin{bmatrix} u_\phi \\ u_z \end{bmatrix} \qquad \begin{bmatrix} \varepsilon_{\phi\phi} \\ \varepsilon_{\theta\theta} \end{bmatrix} \qquad \begin{bmatrix} n_{\phi\phi} \\ n_{\theta\theta} \end{bmatrix} \qquad \begin{bmatrix} p_\phi \\ p_z \end{bmatrix}$$

Displacements   Strains   Membrane forces   Loads

Kinematic   Constitutive   Equilibrium
relation   relation   relation

Fig. 14.3 Scheme of relationships

$$\mathbf{u} = \begin{bmatrix} u_\phi & u_z \end{bmatrix}^T$$

$$\mathbf{e} = \begin{bmatrix} \varepsilon_{\phi\phi} & \varepsilon_{\theta\theta} \end{bmatrix}^T$$

$$\mathbf{s} = \begin{bmatrix} n_{\phi\phi} & n_{\theta\theta} \end{bmatrix}^T$$

$$\mathbf{p} = \begin{bmatrix} p_\phi & p_z \end{bmatrix}^T$$

The relation between these vectors is presented in Fig. 14.3.

14.2 Kinematic Relation

In Chap. 2, we found for the strain ε_{xx} in the system of axes (x, y, z)

$$\varepsilon_{xx} = \frac{\partial u_x}{dx} + \frac{u_z}{r_1} \tag{14.1}$$

Comparing the coordinate system (ϕ, θ, z) of Fig. 14.1b with the (x, y, z) co-ordinate system of Fig. 14.1a, we observe that an infinitesimal increase $d\phi$ of coordinate ϕ causes an increase dx of coordinate x, so $dx = r_1 d\phi$. Therefore, Eq. (14.1) is in the coordinate system (ϕ, θ, z):

$$\varepsilon_{\phi\phi} = \frac{\partial u_\phi}{r_1 d\phi} + \frac{u_z}{r_1} \tag{14.2}$$

For the strain ε_{yy} in y-direction (here strain $\varepsilon_{\theta\theta}$), we found in Chap. 2 a contribution to the displacement u_z in normal direction of the size u_z/r_2. Note that the displacement u_z and radius r_2 are in the same direction. The contribution to the strain $\varepsilon_{\theta\theta}$ is due to the fact that radius r_2 increases to $r_2 + u_z$. In the present chapter, the strain $\varepsilon_{\theta\theta}$ also increases due to the displacement u_ϕ in the direction of the meridian. If a latitude circle displaces over the distance u_ϕ along the tangent to the meridian, the additional circumferential strain in the latitude circle is u_ϕ/r_t, where r_t is defined in Fig. 14.4. It is the distance from point P to the axis of rotation, measured along the tangent line to the shell. The proof is straightforward. The circumference of the latitude circle without the displacement is $u_\phi \cos \phi$ $2\pi r = 2\pi(r_t \cos \phi)$. With the displacement the circumference increases to

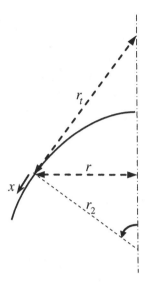

Fig. 14.4 Definition of tangential radius r_t

$2\pi\big((r_t + u_\phi)\cos\phi\big)$. The difference is $2\pi\big(u_\phi\cos\phi\big)$. If we divide by the initial circumference $2\pi(r_t\cos\phi)$, we obtain u_ϕ/r_t, q.e.d. The final result for $\varepsilon_{\theta\theta}$ is

$$\varepsilon_{\theta\theta} = \frac{u_\phi}{r_t} + \frac{u_z}{r_2}. \tag{14.3}$$

Equations (14.2) and (14.3) can symbolically be presented as the kinematic relation $\mathbf{e} = \mathbf{B}\mathbf{u}$:

$$\begin{bmatrix} \varepsilon_{\phi\phi} \\ \varepsilon_{\theta\theta} \end{bmatrix} = \begin{bmatrix} \dfrac{1}{r_1}\dfrac{d}{d\phi} & \dfrac{1}{r_1} \\ \dfrac{1}{r_t} & \dfrac{1}{r_2} \end{bmatrix} \begin{bmatrix} u_\phi \\ u_z \end{bmatrix} \tag{14.4}$$

14.3 Constitutive Relation

The constitutive relation is written as:

$$\begin{bmatrix} n_{\phi\phi} \\ n_{\theta\theta} \end{bmatrix} = \frac{Et}{1 - \nu^2} \begin{bmatrix} 1 & \nu \\ \nu & 1 \end{bmatrix} \begin{bmatrix} \varepsilon_{\phi\phi} \\ \varepsilon_{\theta\theta} \end{bmatrix} \tag{14.5}$$

If the vector $\mathbf{s}$ of shell forces is directly known, the strains can be computed from the inverse relation for the deformation vector $\mathbf{e} = \mathbf{D}^{-1}\mathbf{s}$, which reads:

$$\begin{bmatrix} \varepsilon_{\phi\phi} \\ \varepsilon_{\theta\theta} \end{bmatrix} = \frac{1}{Et} \begin{bmatrix} 1 & -\nu \\ -\nu & 1 \end{bmatrix} \begin{bmatrix} n_{\phi\phi} \\ n_{\theta\theta} \end{bmatrix} \tag{14.6}$$

14.4 Equilibrium Relation

Consider an infinitesimal shell element of size $r_1 d\phi$ in meridian direction and $rd\theta$ in circumferential direction as shown in Fig. 14.5. We derive two equilibrium equations, one for the meridional and one for the normal direction. As shown in Fig. 14.5, two membrane forces play a role and two load components are applied; the system is statically determinate. The edge of the small shell element with size $r_1 d\phi$ is loaded by the membrane force $n_{\theta\theta} \cdot r_1 d\phi$, and the edge with size $rd\theta$ with the force $n_{\phi\phi} \cdot rd\theta$. The latter force increases in ϕ-direction, and the increase points in the ϕ-direction, see Fig. 14.6. Furthermore, the membrane forces $n_{\phi\phi} \cdot rd\theta$ include a small angle $d\phi$ and have a resultant force $n_{\phi\phi} \cdot rd\theta \cdot d\phi$ in normal direction, pointing in the opposite direction of the load $p_z \cdot rd\theta \cdot r_1 d\phi$ on the small shell element. The two circumferential forces $n_{\theta\theta} \cdot r_1 d\phi$ include a small angle $d\theta$, and have a resultant $n_{\theta\theta} \cdot r_1 d\phi \cdot d\theta \cdot \cos\phi$ parallel to the tangent to the meridian and pointing in the direction of decreasing ϕ (see Fig. 14.7). The forces $n_{\theta\theta} \cdot r_1 d\phi$ have another resultant $n_{\theta\theta} \cdot r_1 d\phi \cdot d\theta \cdot \sin\phi$ normal to the meridian,

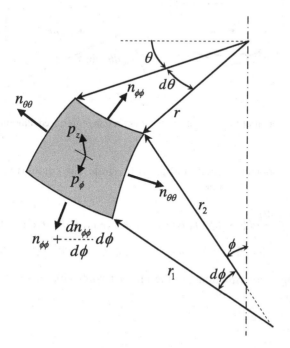

Fig. 14.5 Load components and stress resultants on an infinitesimal element

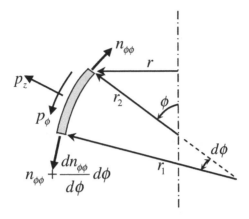

Fig. 14.6 Contribution of the meridional stress resultant to the equilibrium

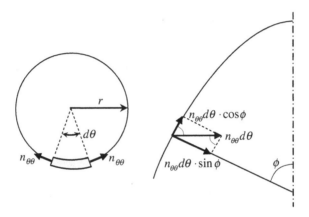

Fig. 14.7 Contribution of the circumferential stress resultant to the equilibrium

pointing in the opposite direction of $p_z \cdot rd\theta \cdot r_1 d\phi$. The equilibrium equations for the membrane forces become

$$\frac{d\left(n_{\phi\phi} \cdot rd\theta\right)}{d\phi}d\phi - n_{\theta\theta} \cdot r_1 d\phi \cdot d\theta \cdot \cos\phi + p_\phi \cdot rd\theta \cdot r_1 d\phi = 0$$

$$-n_{\theta\theta} \cdot r_1 d\phi \cdot d\theta \cdot \sin\phi - n_{\phi\phi} \cdot rd\theta \cdot d\phi + p_z \cdot rd\theta \cdot r_1 d\phi = 0. \tag{14.7}$$

By dividing these equations by the product $r_1 d\phi \cdot d\theta$, we obtain two equilibrium equations

$$\frac{d(n_{\phi\phi}r)}{r_1 d\phi} - n_{\theta\theta}\cos\phi + p_\phi r = 0$$

$$-\frac{n_{\phi\phi}r}{r_1} - n_{\theta\theta}\sin\phi + p_z r = 0. \tag{14.8}$$

Since $r = r_2\sin\phi = r_t\cos\phi$, the equilibrium equations can be rewritten as

$$\frac{1}{r_1}\frac{d(n_{\phi\phi}r)}{d\phi} - \frac{n_{\theta\theta}r}{r_t} + p_\phi r = 0$$

$$-\frac{n_{\phi\phi}r}{r_1} - \frac{n_{\theta\theta}r}{r_2} + p_z r = 0. \tag{14.9}$$

Symbolically presented as $\mathbf{B}^*(\mathbf{s}\,r) = \mathbf{p}\,r$, this equilibrium relation is

$$\begin{bmatrix} -\dfrac{1}{r_1}\dfrac{d}{d\phi} & \dfrac{1}{r_t} \\ \dfrac{1}{r_1} & \dfrac{1}{r_2} \end{bmatrix} \begin{bmatrix} n_{\phi\phi}r \\ n_{\theta\theta}r \end{bmatrix} = \begin{bmatrix} p_\phi r \\ p_z r \end{bmatrix} \tag{14.10}$$

The last equation in Eq. (14.10) can be rewritten as

$$-\frac{n_{\phi\phi}}{r_1} - \frac{n_{\theta\theta}}{r_2} + p_z = 0 \tag{14.11}$$

According to the definition in Eqs. (2.7) and (2.8) we can express the curvature of the shell with the geometry of Fig. 14.5 by

$$k_\phi = -\frac{1}{r_1}, \; k_\theta = -\frac{1}{r_2} \tag{14.12}$$

Using these expressions, we rewrite the equilibrium equation:

$$k_\phi n_{\phi\phi} + k_\theta n_{\theta\theta} + p_z = 0 \tag{14.13}$$

This equation is similar to the last equilibrium equation in Eq. (2.19) in Chap. 2.

14.5 Membrane Forces and Displacements

14.5.1 Membrane Forces

Because the system of equilibrium equations is statically determinate, the membrane forces can be obtained directly from the equilibrium equations. The circumferential membrane force $n_{\theta\theta}$ is eliminated from the equilibrium relation (14.10) by multiplying the first equation with $\sin\phi$ and the second with $\cos\phi$:

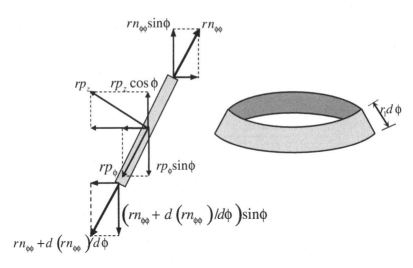

Fig. 14.8 Vertical equilibrium of a shell ring

$$\frac{d(n_{\phi\phi}r)}{r_1 d\phi} \sin\phi + \frac{n_{\phi\phi}r}{r_1}\cos\phi = -(p_z \cos\phi - p_\phi \sin\phi)r \qquad (14.14)$$

The left-hand member of this equation can be written as $d(n_{\phi\phi}r\sin\phi)/r_1 d\phi$, so the equation is also given by:

$$\frac{d(n_{\phi\phi}r\sin\phi)}{r_1 d\phi} = -(p_z \cos\phi - p_\phi \sin\phi)r \qquad (14.15)$$

This new equilibrium condition has a physical meaning. It describes the vertical equilibrium of a shell ring between two adjacent parallel circles with radius r at distance $d\phi$ from each other, see Fig. 14.8. Equilibrium of the shell ring in the direction of the axis of rotation requires

$$\frac{d(n_{\phi\phi}r\ \sin\phi)}{d\phi}d\phi + (p_z \cos\phi - p_\phi \sin\phi)r\ r_1 d\phi = 0 \qquad (14.16)$$

In this equation, we have included the forces which act on a ring part over an angle θ of one radial. If we divide by $r_1\,d\phi$, we obtain Eq. (14.15). Integration of this equation gives the meridional membrane force

$$n_{\phi\phi} = \frac{1}{r\sin\phi}\int (p_z \cos\phi - p_\phi \sin\phi)rr_1 d\phi. \qquad (14.17)$$

Substituting this result in the second equilibrium equation of Eq. (14.10) leads to

$$n_{\theta\theta}r_1 \sin\phi = p_z rr_1 - \frac{1}{\sin\phi}\int (p_z \cos\phi - p_\phi \sin\phi)rr_1 d\phi. \qquad (14.18a)$$

Dividing by $r_1 \sin \phi$, and accounting for $r_2 = r/\sin \phi$, we rewrite the equation:

$$n_{\theta\theta} = p_z r_2 - \frac{1}{r_1 \sin^2 \phi} \int (p_z \cos \phi - p_\phi \sin \phi) r r_1 d\phi. \qquad (14.18b)$$

We end up with the membrane solution:

$$n_{\phi\phi} = \frac{1}{r \sin \phi} F(\phi)$$

$$n_{\theta\theta} = p_z r_2 - \frac{1}{\sin^2 \phi} \frac{F(\phi)}{r_1} \qquad (14.19)$$

The term $F(\phi)$ in Eq. (14.19) is the expression for the integrated load terms:

$$F(\phi) = \int f(\phi) d\phi = \int (p_z \cos \phi - p_\phi \sin \phi) r r_1 d\phi. \qquad (14.20)$$

14.5.2 Displacements

The shells of revolution, considered in this chapter, are loaded by axisymmetric loads. The only place where displacements are of true interest is at the base of the shell. There, dependent on the type of support, disturbance of the displacement field may occur, and bending moments may come into being in an edge zone. This will be the subject of the next chapter. Then it is necessary to know the displacements at the base due to the membrane forces. These are the radial displacement u_r in the horizontal plane and the rotation φ_ϕ, defined in Fig. 14.9. We find displacements by integrating strains, and the strains are obtained by the constitutive relations (14.6):

$$\varepsilon_{\phi\phi} = \frac{1}{Et} (n_{\phi\phi} - \nu n_{\theta\theta})$$

$$\varepsilon_{\theta\theta} = \frac{1}{Et} (-\nu n_{\phi\phi} + n_{\theta\theta}) \qquad (14.21)$$

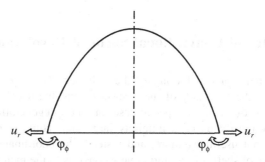

Fig. 14.9 Definition of displacement and rotation at the base

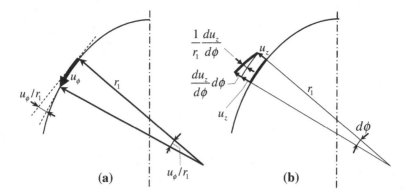

Fig. 14.10 Rotation of the tangent to the meridian: **a** due to u_ϕ, **b** due to u_r

The radial displacement is easily calculated from the circumferential strain at the base:

$$u_r = \varepsilon_{\theta\theta} r_{base}.$$ (14.22)

A rotation φ_ϕ is due to both the displacement u_ϕ and the displacement u_z, as illustrated in Fig. 14.10. The meridional displacement u_ϕ causes a change of angle of the tangent line at point P of the size u_ϕ/r_1. Due to the normal displacement u_z, the tangent line rotates over an angle $-du_z/(r_1 d\phi)$. The two contributions together lead to the equation for the rotation φ_ϕ:

$$\varphi_\phi = \frac{1}{r_1}\left(u_\phi - \frac{du_z}{d\phi}\right).$$ (14.23)

At the base of the shell of revolution, the meridional displacement u_ϕ will be zero. Then the equation simplifies to

$$\varphi_\phi = -\frac{du_z}{r_1 d\phi}.$$ (14.24)

14.6 Geometry of Conventional Shells of Revolution

Often applied shells of revolution are the cylinder, the cone and the sphere. Fig. 14.11 shows the geometry of those surfaces and the usual co-ordinate systems. The circular cylinder is a special case, in fact a degeneration, of a shell of revolution. Now the radius r_1 is infinitely large and radius r_2 is constant. The coordinate ϕ is not useful anymore, and instead, the coordinate x is used. The membrane theory of shells of revolution now converts to the membrane theory for

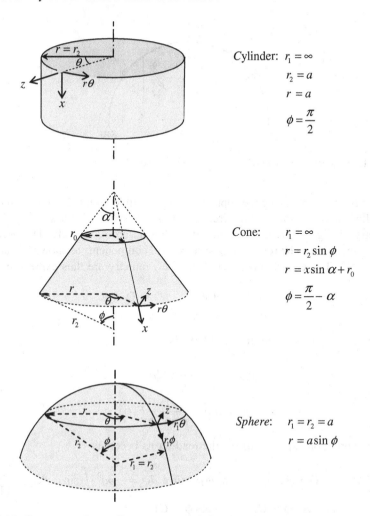

Cylinder: $r_1 = \infty$

$r_2 = a$

$r = a$

$\phi = \dfrac{\pi}{2}$

Cone: $r_1 = \infty$

$r = r_2 \sin \phi$

$r = x \sin \alpha + r_0$

$\phi = \dfrac{\pi}{2} - \alpha$

Sphere: $r_1 = r_2 = a$

$r = a \sin \phi$

Fig. 14.11 Geometry and co-ordinate system of shells of revolution

circular cylinders as discussed in Chap. 4. For an application we refer to the water tank example of Sect. 4.9.

14.7 Application to a Spherical Shell Under its Own Weight

Consider a spherical shell subject to its own weight p, supported by rollers at the base, see Fig. 14.11. The membrane response of the shell leads to a meridional membrane force $n_{\phi\phi}$ and a circumferential membrane force $n_{\theta\theta}$. A pure membrane

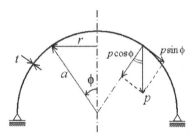

Fig. 14.12 Geometry and load of a roller supported sphere

solution is applicable since the supports are compatible with membrane forces and they allow deformation in the circumferential direction. For a sphere with the geometry of Fig. 14.12, the meridian describes an arc of a circle. The vertical's own weight can thus be rewritten to surface load components as shown in Sect. 5.6 and Fig. 5.3. These load components and the geometry are thus expressed by

$$p_\phi = p \, \sin \phi \qquad r_1 = r_2 = a$$
$$p_z = -p \, \cos \phi \qquad r = a \, \sin \phi \tag{14.25}$$

The membrane solution in Eq. (14.19) is

$$n_{\phi\phi} = \frac{1}{a \sin^2 \phi} F(\phi)$$
$$n_{\theta\theta} = p_z a - \frac{1}{a \sin^2 \phi} F(\phi) \tag{14.26}$$

The expression Eq. (14.20) for the load terms becomes

$$F(\phi) = \int \left(-p \cos^2 \phi - p \sin^2 \phi \right) a^2 \sin \phi \, d\phi = -pa^2 \int \sin \phi \, d\phi$$
$$= pa^2 (\cos \phi + C) = pa^2 (\cos \phi + C). \tag{14.27}$$

The membrane forces are consequently

$$n_{\phi\phi} = pa \left(\frac{\cos \phi + C}{\sin^2 \phi} \right)$$
$$n_{\theta\theta} = -pa \left(\cos \phi + \frac{\cos \phi + C}{\sin^2 \phi} \right) \tag{14.28}$$

We must determine the constant C, but we know neither $n_{\phi\phi}$ nor $n_{\theta\theta}$. We can get round this obstacle by rewriting the expression for $n_{\phi\phi}$ in Eq. (14.28):

$$n_{\phi\phi} \sin^2 \phi = pa(\cos \phi + C) \tag{14.29}$$

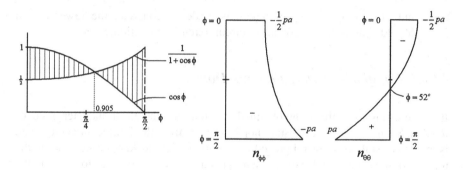

Fig. 14.13 Distribution of the stress resultants over the sphere

If we now substitute $\phi = 0$, we obtain constant C:

$$0 = pa(1 + C) \quad \Rightarrow \quad C = -1. \tag{14.30}$$

We could have determined C also in an alternate way by considering the stress state at the top (apex) of the sphere. There, the membrane forces $n_{\phi\phi}$ and $n_{\theta\theta}$ are equal, because each meridian crosses another meridian perpendicularly. If we substitute $n_{\phi\phi} = n_{\theta\theta}$ and $r_1 = r_2 = a$ into equilibrium Eq. (14.10), we find $n_{\phi\phi} = n_{\theta\theta} = -pa/2$. From this solution, we can solve the constant by substituting $\phi = 0$ in Eq. (14.28). This also yields $C = -1$.

Substituting the constant into the expressions for the membrane stress resultants and rewriting this result by using the relation $\sin^2 \phi = 1 - \cos^2 \phi$ gives:

$$
\begin{aligned}
n_{\phi\phi} &= -pa\frac{1}{1 + \cos \phi} \\
n_{\theta\theta} &= pa\left(\frac{1}{1 + \cos \phi} - \cos \phi\right)
\end{aligned} \tag{14.31}
$$

Their own weight leads to a compressive meridional membrane force for every point of the shell. The circumferential membrane force however is partly compressive and partly tensile depending on the position of the parallel circle, denoted by the angular co-ordinate ϕ. The course and value of the membrane forces are shown in Fig. 14.13 between the top ($\phi = 0$) and the edge ($\phi = \pi/2$) of the sphere. Of interest to the designer is the position for which the circumferential stress resultant is equal to zero. If we substitute for reasons of ease $\cos \phi = x$, we must solve

$$\frac{1}{1 + x} - x = 0 \quad \rightarrow \quad 1 - x - x^2 = 0. \tag{14.32}$$

The solution is $x = 0.618$, therefore $\phi = 0.905$ rad, about $52°$. In Fig. 14.13, the solution has been done also in a graphical way by intersecting the functions

$\cos\phi$ and $(1 + \cos\phi)^{-1}$. If the sphere is made of concrete, the lower region $52° \leq \phi \leq 90°$ has to be reinforced in circumferential direction.

14.7.1 Comparison with Famous Domes

It is interesting to confront the obtained solution with the 35 m diameter dome of the Aya Sophia in former Constantinople. It dates from the fourteenth century and is the final result after several preceding collapses. The thickness is constant in the top part of the sphere, and starts to increase at an angle ϕ very close to 52°. Did the rebuilders understand that tensile stresses occur in the bottom part of the sphere and that the tensile stress of brick material is less than compressive strength? And were they able to make an estimate on the size of the tensile stress and the zone in which it occurs? A similar interesting observation can be made regarding the Saint Peter dome in Rome. Also here the thickness is constant in the top part and increases in the direction of the base in the bottom part.

14.7.2 Approximation as Short Beam

We can make the result of the membrane theory accessible by considering the equilibrium of a half sphere as shown in Fig. 14.14 and calculating the bending moment in the plane of symmetry. In fact we consider the doem as a shot beam. In a standard mathematical exercise we can calculate that the centre of gravity of the half sphere is at a distance $4a/\pi^2$ from the plane of symmetry. The total weight F of the half sphere is $2pa^2$. The vertical support reaction is uniformly distributed over the half base circle, and the resulting support reaction F is at a distance $2a/\pi$ to the plane of symmetry. The bending moment in this plane is $M = eF$, or

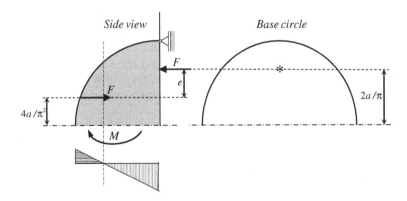

Fig. 14.14 Rough check on the sphere membrane solution

$M = 2pa^2(2a/\pi - 4a/\pi^2)$, or $M = (4/\pi - 8/\pi^2)\,pa^3$. Worked out, the moment is $M = 0.463\,pa^3$. We can make a first guess of the order of magnitude of the tensile membrane force $n_{\theta\theta}$ if we assume a linear distribution over the cross-section of the sphere in the plane of symmetry. This cross-section is a half circle, the horizontal neutral line is at a distance $2a/\pi$ from the base, and the moment of resistance to calculate the tensile membrane force at the base from the bending moment is $W = (\pi^2/4 - 2)a^2$, which worked out is $W = 0.467a^2$. Then the membrane force at the base is $n_{\theta\theta} = M/W$, or $n_{\theta\theta} = 0.463/0.467\,pa$. We find $n_{\theta\theta} \approx pa$ which is in amazing agreement with the exact value in Fig. 14.13.

14.8 Application to a Conical Shell Subject to its Own Weight

Consider a conical shell subject to its own weight p (see Fig. 14.15). This conical shell is supported by a single column at the top. The membrane response of the shell leads to a meridional membrane fore $n_{\phi\phi}$ and a circumferential membrane force $n_{\theta\theta}$. The determination of these membrane forces will give a broad insight into the response of the conical shell. The geometry is shown in Fig. 14.15; the meridian is a straight line. The vertical's own weight can thus be reduced to surface load components, as shown in Fig. 14.12. The angle ϕ is constant and therefore the load components are also constant. The load components are

$$p_\phi = p \sin \phi$$
$$p_z = -p \cos \phi \tag{14.33}$$

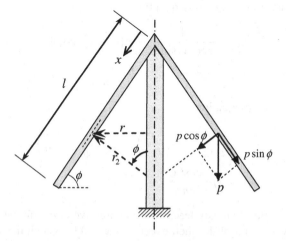

Fig. 14.15 Meridional geometry and load of conical shell on single column

Since the angle ϕ is constant, it is no longer a co-ordinate and the corresponding radius of curvature r_1 is infinite. Therefore the co-ordinate x is introduced, which is measured from the top along the meridian. To make the membrane solution applicable to a conical shell, the following observations are made for the meridional direction and the radius of the parallel circle:

$$
\left.\begin{aligned}
r_1 &= \infty \\
\phi &= constant
\end{aligned}\right\} \quad r_1 d\phi = dx
$$

$$
\left.\begin{aligned}
x/r_2 = \tan\phi \quad &\Rightarrow \quad r_2 = x\cot\phi \\
r/r_2 = \sin\phi \quad &\Rightarrow \quad r = r_2\sin\phi
\end{aligned}\right\} \quad r = x\cos\phi.
$$

(14.34)

The membrane solution of Eq. (14.19) is

$$
n_{xx} = \frac{1}{x\cos\phi\sin\phi}F(x)
$$

$$
n_{\theta\theta} = p_z r_2
$$

(14.35)

Equation (14.20) for the load terms becomes:

$$
F(x) = \int_0^x \left[(-p\cos^2\phi - p\sin^2\phi)x\cos\phi\right]dx
$$

$$
= -p\cos\phi \cdot \frac{1}{2}x^2 + C
$$

(14.36)

At the free edge, the meridional membrane force must be zero:

$$
x = l; \quad n_{xx} = 0 \quad \Rightarrow \quad F(l) = 0 \quad \Rightarrow \quad C = \frac{1}{2}pl^2\cos\phi.
$$

(14.37)

The membrane forces are consequently

$$
n_{xx} = \frac{1}{x\cos\phi\sin\phi} \cdot \frac{1}{2}p(l^2 - x^2)\cos\phi
$$

$$
n_{\theta\theta} = p_z r_2 = -p\cos\phi \cdot \frac{x}{\tan\phi}.
$$

(14.38)

We rearrange these equations:

$$
n_{xx} = \frac{pl}{2\sin\phi}\left(\frac{l^2 - x^2}{xl}\right) = \frac{pl}{2\sin\phi}\left(\frac{l}{x} - \frac{x}{l}\right)
$$

$$
n_{\theta\theta} = -px\frac{\cos^2\phi}{\sin\phi}.
$$

(14.39)

Their own weight obviously leads to a compressive circumferential membrane force $n_{\theta\theta}$ for every point of the shell except for $x = 0$ for which the stress is equal to zero. This membrane force increases linearly from top to bottom. The

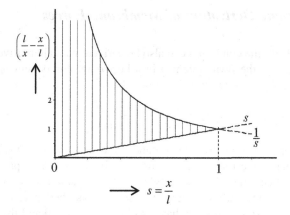

Fig. 14.16 The factor $\left(\frac{l}{x} - \frac{x}{l}\right)$ for $0 \le \frac{x}{l} \le 1$

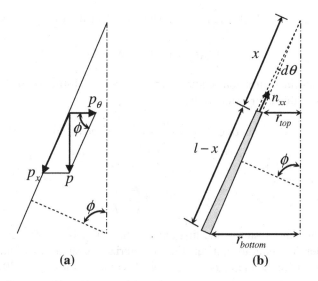

Fig. 14.17 a Decomposition of own weight p. **b** Equilibrium in x-direction

meridional membrane force $n_{\phi\phi}$ becomes infinite at the top and is tensile for every point except for the free edge where the stress is equal to zero. A representation of this membrane force is shown in Fig. 14.16, where the course of the force is presented graphically by the shaded area.

14.8.1 Alternate Derivation of Membrane Forces

The Eq. (14.39) for the cone shell can also be derived in an alternative, easier way. Decompose their in the own weight p in a horizontal component p_θ and a component p_x along the shell as shown in Fig. 14.17a:

$$p_x = \frac{p}{\sin \phi}; \quad p_\theta = \frac{p}{\tan \phi}. \tag{14.40}$$

Component p_x causes tensile membrane forces n_{xx} and component p_θ causes compressive membrane forces $n_{\theta\theta}$. Figure 14.17b shows a shell part enclosed by meridians in vertical planes with an angle $d\theta$. It starts at a distance x from the top and continues to the bottom edge of the shell. This shell part is a symmetric trapezium. The horizontal top side has a width $w_{top} = r_{top}d\theta$ and the bottom edge has a width $w_{bottom} = r_{bottom}d\theta$:

$$w_{top} = x \cos \phi \, d\theta; \quad w_{bottom} = l \cos \phi \, d\theta. \tag{14.41}$$

The length of the shell part is $l - x$. The area A of the trapezium is

$$A = \frac{1}{2}(l - x)(l + x) \cos \phi \, d\theta. \tag{14.42}$$

Equilibrium in x-direction requires $n_{xx}w_{top} = p_x A$. Substitution of Eqs. (14.41) and (14.42) in this expression yields

$$n_{xx}x \cos \phi \, d\theta = \frac{p}{2 \sin \phi} \left(l^2 - x^2\right) \cos \phi \, d\theta \tag{14.43}$$

from which we obtain

$$n_{xx} = \frac{pl}{2 \sin \phi} \left(\frac{l}{x} - \frac{x}{l}\right). \tag{14.44}$$

This is the same result as the first expression in Eq. 14.39. The calculation of $n_{\theta\theta}$ is simpler. The equilibrium equation of a horizontal shell ring of unit length is $n_{\theta\theta} = -p_\theta r$, where the radius is $r = x \cos \phi$. Accounting for Eq. (14.40) we obtain

$$n_{\theta\theta} = -px \frac{\cos^2 \phi}{\sin \phi}. \tag{14.45}$$

which is the second expression in Eq. (14.39).

Chapter 15
Edge Disturbance in Shell of Revolution Due to Axisymmetric Loading

Tens of years ago, theories were proposed for the edge disturbance problem in shells of revolution. Because a rigorous bending theory is complicated, attempts were made for reliable approximations. A well-known one was published by Geckeler [1, 2], who obtained his approximation by simplifying mathematical considerations to the exact equations. We adopt in the present chapter the approximation of Geckeler, but will arrive at it in an alternative, simpler way, which is inspired by the stave-ring model for edge disturbances in circular cylindrical shells in Chap. 5.

15.1 Problem Statement

The membrane solution will be used as an inhomogeneous solution, so we only have to find a homogeneous solution for the edge disturbance. We focus on the four different shell geometries shown in Fig. 15.1: the circular cylindrical, the conical and two spherical shells; one sphere has a cylinder as envelope and one a cone. The envelope is the tangent plane at the base circle. In Chap. 5, we learned that the influence length of the edge disturbance in the cylindrical shell is about one quarter of the radius of the cylinder. The disturbance really occurs in a small edge zone. This means that we can use the solution of the cylinder (a) in Fig. 15.1 also for the sphere (c); at the edge, the cylinder is the tangent plane of the sphere, and the difference between the edge zone of sphere and cylinder is negligible. For the cone (b) we will have another solution than for the cylinder, but again we expect a disturbance that is limited to an edge zone. If we have solved the cone problem, we can also use this solution for the sphere (d). Therefore, we will focus in the present chapter on the cone problem.

J. Blaauwendraad and J. H. Hoefakker, *Structural Shell Analysis*, Solid Mechanics and Its Applications 200, DOI: 10.1007/978-94-007-6701-0_15, © Springer Science+Business Media Dordrecht 2014

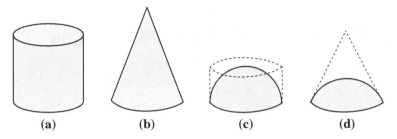

Fig. 15.1 Four examples of shell of revolution

15.2 Recall of Solution for Circular Cylinder

We recall the main findings from Chap. 5, where we solved the problem as depicted in the top half of Fig. 15.2. This solution is also applicable to the system depicted in the bottom half of the figure.

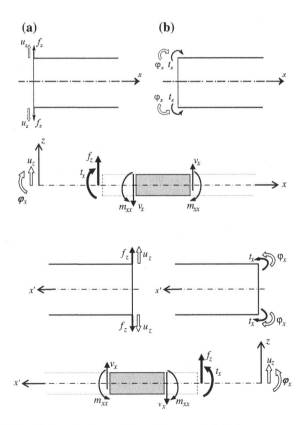

Fig. 15.2 Definition of edge displacements and forces, and shell forces

We considered the circular cylinder as a barrel of staves and rings, where the rings behave as elastic supports to the staves. In Fig. 15.2, we show the sign convention for the bending moment m_{xx} and transverse shear force v_x in the staves. We derived the differential equation

$$D\frac{d^4 u_z}{dx^4} + k_z u_z = 0; \quad k_z = \frac{Et}{a^2} \tag{15.1}$$

where D is the flexural rigidity of the staves, including the effect of Poisson's ratio, and k_z the spring stiffness of the rings. We introduced the parameter β:

$$\beta^4 = \frac{Et}{4Da^2} = \frac{3(1 - v^2)}{(at)^2} \tag{15.2}$$

and obtained the differential equation

$$\frac{d^4 u_z}{dx^4} + 4\beta^4 u_z = 0 \tag{15.3}$$

For shells longer than the influence length, an edge disturbance has attenuated before it reaches the opposite edge. Therefore we used the solution

$$u_z(x) = Ce^{-\beta x} \sin(\beta x + \psi) \tag{15.4}$$

This leads to bending moments and transverse shear forces in the staves and circumferential membrane forces in the rings given by:

$$
\begin{aligned}
m_{xx} &= -D\left(2\beta^2 Ce^{-\beta x} \sin(\beta x + \psi - \pi/2)\right) \\
v_x &= -D\left(-2\sqrt{2}\beta^3 Ce^{-\beta x} \sin(\beta x + \psi - 3\pi/4)\right) \\
n_{\theta\theta} &= \frac{Et}{a} Ce^{-\beta x} \sin(\beta x + \psi)
\end{aligned}
\tag{15.5}
$$

The displacement and rotation at the shell end $x = 0$ are

$$
\begin{aligned}
u_z &= C \sin \psi \\
\varphi_x &= \sqrt{2}\, \beta\, C\, \sin(\psi - \pi/4)
\end{aligned}
\tag{15.6}
$$

On the basis of this theory, we have derived the four elementary cases of Figs. 5.12 and 5.13 and Table 5.1. We also derived the *flexibility matrix equation* for forces and displacements shown in Fig. 15.2:

$$\frac{1}{D_b}\begin{bmatrix} \dfrac{1}{2\beta^3} & \dfrac{1}{2\beta^2} \\ \dfrac{1}{2\beta^2} & \dfrac{1}{\beta} \end{bmatrix}\begin{bmatrix} f_z \\ t_x \end{bmatrix} = \begin{bmatrix} u_z \\ \varphi_x \end{bmatrix} \tag{15.7}$$

and a formula for the characteristic length l_c and the influence length l_i

$$l_c = \frac{\sqrt{at}}{\sqrt[4]{3(1-v^2)}}; \quad l_i = \pi l_c \approx 2.5\sqrt{at} . \tag{15.8}$$

15.3 Extension to Cones

Consider the cone of Fig. 15.3a with base circle of radius a. We choose an x-ordinate along the straight meridian, starting at the cone top. We will use the concept of a barrel with staves and rings again. The staves are shown in Fig.15.3b and the rings in Fig.15.3c. There are a number of differences with the application to circular cylinders:

1. The staves are not prismatic but tapered.
2. The radius of the rings is not constant.
3. In cylinders the membrane force n_{xx} in axial direction is zero, but not in cones.
4. The displacement u_z is in the direction of the ring radius for cylinders, but not for cones.

The items 1 and 2 are not a real hindrance. The disturbance remains limited to a narrow zone near the edge, therefore the effect of tapering will be very small, and the change of radius can be neglected. We can assume prismatic staves, and rings of equal length. For rings we use the radius a of the base. Item 3, the existence of nonzero membrane forces in x-direction, implies strains ε_{xx} and therefore nonzero displacements u_ϕ in x-direction

Consider a horizontal ring load in radial direction at the base of the cone, see the left part of Fig. 15.4; it will cause both axial membrane deformation and lateral bending deformation, which is expected to occur in a zone with a certain influence length. We do not yet know this influence length l_i, but we expect it to be related to $\sqrt{at}$, therefore short. The axial membrane deformation is of the order l_i/Et, and the bending deformation is of the order l_i^3/Et^3. The ratio is t^2/l_i^2, which is more or less the ratio of u_ϕ and u_z. Because l_i is related to $\sqrt{at}$, the ratio is t/a; therefore we

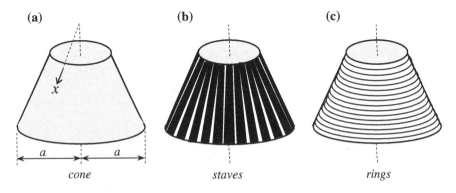

(a)   **(b)**   **(c)**

cone   staves   rings

Fig. 15.3 **a** Chosen x-axis in cone, **b** staves, and **c** rings

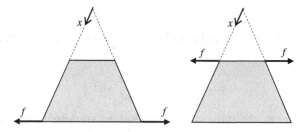

Fig. 15.4 Cones with horizontal ring force

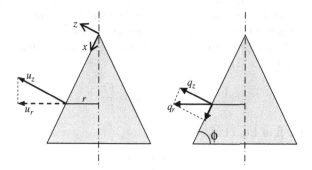

Fig. 15.5 Ring quantities u_r and q_r

conclude that the displacement u_ϕ is an order of magnitude smaller than the normal displacements u_z, and we set u_ϕ equal to zero.

Consider the cone of Fig. 15.5. As we did in Chap. 5, we introduce the angle ϕ and the interaction force q_r between the staves and the rings. In Chap. 5, we called the force q_z because it is directed in the normal direction z; now it does not. Consider a point P at a distance x from the edge. At this position we call the radius of the cone in the horizontal plane r. The radial force q_r acts in positive r-direction on the rings and in negative r-direction on the stave. The displacement of P consists of a normal displacement u_z only. The new position of the point is P′. The increase u_r of the radius r due to the normal displacement is

$$u_r = u_z \sin \phi \qquad (15.9)$$

The force q_r to produce this displacement is, equivalent to Eq. (5.4),

$$q_r = \frac{Et}{r^2} u_r \qquad (15.10)$$

This force acts on the stave under an angle ϕ. We must decompose it in two components, one normal to the stave (q_z) in negative z-direction and one in its positive x-direction (q_x). The latter equilibrates the support reaction, and the first provides the elastic support of the stave. It holds that

$$q_z = q_r \sin \phi \qquad (15.11)$$

We substitute Eq. (15.10) in Eq. (15.11), accounting for Eq. (15.9) and noticing $r \approx a$,

$$q_z = k_z u_z; \quad k_z = \frac{Et}{a^2} \sin^2 \phi \qquad (15.12)$$

The behaviour of the stave remains unchanged compared to the circular cylinder. We conclude that the parallelism with the 'beam on elastic foundation' remains valid; the only difference with the circular cylinder is the calculation of the spring constant k_z. Now the additional factor $\sin^2 \phi$ comes in, which is unity for $\phi = \pi/2$, the value for a circular cylinder. The 'elastic foundation' of cones is less stiff than of circular cylinders. In fact, the message of Eq. (15.12) is that we must replace radius a by the principal radius r_2. From Fig. 15.6, we obtain $a = r_2 \sin \phi$, which changes Eq. (15.12) into

$$q_z = k_z u_z; \quad k_z = \frac{Et}{r_2^2} \qquad (15.13)$$

If we replace β of Eq. (15.2) by

$$\beta^4 = \frac{3(1 - v^2)}{(r_2 t)^2} \qquad (15.14)$$

then the characteristic length and influence length are

$$l_c = \frac{\sqrt{r_2 t}}{\sqrt[4]{3(1 - v^2)}}; \quad l_i = \pi l_c \approx 2.5 \sqrt{r_2 t} \qquad (15.15)$$

and we can still use Eqs. (15.4) up to (15.6) for circular cylinders. Said in other words, we can replace the cone with base radius a by a circular cylinder with base radius r_2 to obtain the correct differential equation. We show this in two ways in

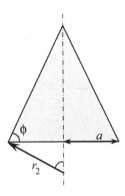

Fig. 15.6 Relation between a and r_2

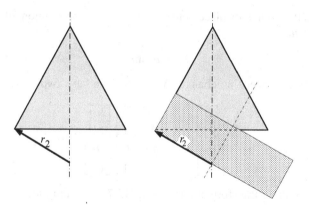

Fig. 15.7 Replacing circular cylinder

Fig. 15.7. In the best way, we use the replacing cylinder in the right part of the figure, which touches the cone such that the walls of the cylinder and cone coincide. For values between 20° and 30°, the result starts to go wrong [3].

As said before, Geckeler arrived at the same result by mathematical considerations. First he derived correct constitutive and equilibrium equations for the rotationally symmetric shell without approximations. After that he introduced mathematical simplifications. Each time when derivatives of a function of different orders appeared, he just kept the highest order derivative and neglected all lower ones. This is permitted if the function varies rapidly, which is the case for edge disturbances. One might say that the derivation, chosen in the present chapter, illustrates the physical background of Geckeler's mathematical approximation.

The last step is to convert Eq. (15.7) to the coordinate system in which the load of Fig. 15.4 fits. For that purpose we refer to Fig. 15.8. Apart from the directions x and z we also define the directions a and r. Ordinate a is in the direction of the axis of revolution, and r is in the direction of the horizontal radius. Decomposition of the force f_r yields f_z of the size

$$f_z = f_r \sin \phi \qquad (15.16)$$

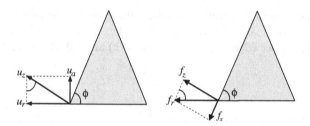

Fig. 15.8 Rotation of axes

Similarly the radial displacement u_r receives a contribution of the normal displacement u_z

$$u_r = u_z \sin \phi \qquad (15.17)$$

The rotation φ_x needs no change. As a result we have two transformations:

$$\begin{bmatrix} f_z \\ t_x \end{bmatrix} = \begin{bmatrix} \sin \phi & 0 \\ 0 & 1 \end{bmatrix} \begin{bmatrix} f_r \\ t_x \end{bmatrix}$$

$$\begin{bmatrix} u_r \\ \varphi_x \end{bmatrix} = \begin{bmatrix} \sin \phi & 0 \\ 0 & 1 \end{bmatrix} \begin{bmatrix} u_z \\ \varphi_x \end{bmatrix} \qquad (15.18)$$

If we apply these transformations to Eq. (15.7) according to

$$\begin{bmatrix} \sin \phi & 0 \\ 0 & 1 \end{bmatrix} \frac{1}{D_b} \begin{bmatrix} \dfrac{1}{2\beta^3} & \dfrac{1}{2\beta^2} \\ \dfrac{1}{2\beta^2} & \dfrac{1}{\beta} \end{bmatrix} \begin{bmatrix} \sin \phi & 0 \\ 0 & 1 \end{bmatrix} \begin{bmatrix} f_r \\ t_x \end{bmatrix} = \begin{bmatrix} u_r \\ \varphi_x \end{bmatrix} \qquad (15.19)$$

we obtain

$$\frac{1}{D_b} \begin{bmatrix} \dfrac{\sin^2 \phi}{2\beta^3} & \dfrac{\sin \phi}{2\beta^2} \\ \dfrac{\sin \phi}{2\beta^2} & \dfrac{1}{\beta} \end{bmatrix} \begin{bmatrix} f_r \\ t_x \end{bmatrix} = \begin{bmatrix} u_r \\ \varphi_x \end{bmatrix} \qquad (15.20)$$

We conclude that the term $\sin \phi$ must be applied two times. It plays a role in the determination of β, and it appears in the flexibility matrix if Eq. (15.20). This latter equation is the flexibility matrix equation for the cone (a) in Fig. 15.9. If the rotations are defined in the opposite direction as shown in (b) of Fig. 15.9, then the matrix equation is

$$\frac{1}{D_b} \begin{bmatrix} \dfrac{\sin^2 \phi}{2\beta^3} & -\dfrac{\sin \phi}{2\beta^2} \\ -\dfrac{\sin \phi}{2\beta^2} & \dfrac{1}{\beta} \end{bmatrix} \begin{bmatrix} f_r \\ t_x \end{bmatrix} = \begin{bmatrix} u_r \\ \varphi_x \end{bmatrix} \qquad (15.21)$$

In accordance with Maxwell's reciprocal theorem, the flexibility matrix is for both cases symmetrical and positive definite. For similar couples of degrees of freedom at the upper edge of a cone the parts (c) and (d) of Fig. 15.9 apply.

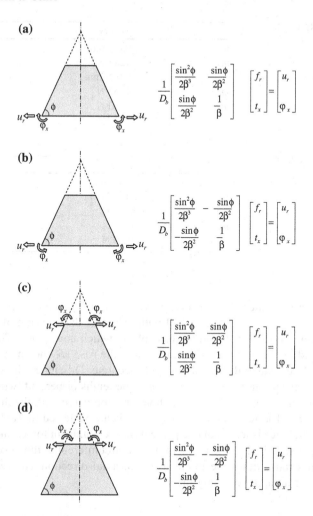

Fig. 15.9 Cone with four different cases of degrees of freedom on edges

15.3.1 Computational Verification

We have checked the accuracy of Geckeler's approximation on the basis of a cone analysis. For this purpose, we have chosen the cone and loading of the left part of Fig. 15.4. We apply Geckeler's theory, and perform a Finite Element analysis. The following data apply: base radius 1000 mm, height of the cone 1000 mm, angle of wall with base plane 60°, shell thickness 10 mm, Young's modulus 2.1×10^5 MPa, Poisson's ratio 0.3, and load at the base 100 N/mm. As for the boundary conditions, the top edge is free; the base edge is free to roll outward horizontally, but cannot rotate. In the FE analysis, we have applied 40 elements over the height of the cone.

Table 15.1 Results for circular cylinder

	Theory	FE analysis
u	0.61 mm	0.61 mm
$n_{\theta\theta}$	1287 N/mm	1288 N/mm
n_{xx}	0.0 N/mm	2 N/mm
$m_{\theta\theta}$	1167 N	1170 N
m_{xx}	3890 N	3890 N

Table 15.2 Results for a cone

	Theory	FE analysis
u	0.57 mm	0.57 mm
$n_{\theta\theta}$	1197 N/mm	1189 N/mm
n_{xx}	50 N/mm	47 N/mm
$m_{\theta\theta}$	1089 N	1080 N
m_{xx}	3620 N	3630 N

In order to guarantee that the FE software is reliable, we first performed the analysis for a circular cylindrical shell with the same data, except of course, the angle between the shell wall and the base plane, which now is 90°. The results of the theory and the FE analysis at the location of the base are shown in Table 15.1. We can firmly conclude that the software is accurate. Differences between theory and FE analysis are on the order of one or some tenths of percent, where we note that the difference in the values of n_{yy} should be compared with the differences in the value of n_{xx}. The results of the cone analysis are collected in Table 15.2. The maximum difference is less than one percent for the dominant forces and moments, $n_{\theta\theta}$ and m_{xx}, respectively. The moment $m_{\theta\theta}$ in circumferential direction is exactly 0.3, Poisson's ratio, times the moment m_{xx} in axial direction The result is very convincing for an approximating theory.

15.4 Application to Clamped Sphere Cap

Consider a spherical shell with clamped edges subject to a uniform normal load $p_z = p$, see Fig. 15.10a. The angle ϕ_o defines the tangent plane at the base circle. In a sphere we use coordinates ϕ and θ. The membrane response of the shell leads to a meridional stress resultant $n_{\phi\phi}$ and a circumferential stress resultant $n_{\theta\theta}$. This membrane solution would be correct if the clamped edges were roller supports. Since the shell cannot deform at the edge circle, a force f_r and torque t_x must work on the edge to satisfy the boundary conditions. This additional bending action decreases along the meridian and evidently represents the edge disturbance. The magnitude of the edge loads is calculated with the aid of Eq. (15.20) between the edge loads and the edge displacements.

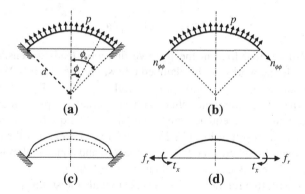

Fig. 15.10 Uniform load (**a**), membrane response (**b**), bending deformation in edge zone (**c**), edge loads (**d**)

15.4.1 Membrane Solution

The membrane solution is given in Eq. (14.19):

$$n_{\phi\phi} = \frac{1}{a \sin^2 \phi} F(\phi)$$

$$n_{\theta\theta} = pa - \frac{1}{a \sin^2 \phi} F(\phi) \tag{15.22}$$

Accounting for Eq. (14.20), the load term is

$$F(\phi) = \int_0^\phi \left[(p \cos \phi) a^2 \sin \phi \right] d\phi = pa^2 \int_0^\phi \sin \phi \, d(\sin \phi)$$

$$= \frac{1}{2} pa^2 \left[\sin^2 \phi \right]_0^\phi + C = \frac{1}{2} pa^2 \sin^2 \phi + C \tag{15.23}$$

The membrane forces are

$$n_{\phi\phi} = \frac{1}{2} pa + \frac{C}{a \sin^2 \phi}$$

$$n_{\theta\theta} = \frac{1}{2} pa - \frac{C}{a \sin^2 \phi} \tag{15.24}$$

There are two considerations to choose the integration constant C zero. First, the term with the constant becomes infinite for $\phi = 0$ if the constant C is nonzero. Secondly, each meridian crosses another meridian perpendicularly at the apex and since the parallel circles are perpendicular to the meridians, the meridional membrane force and the circumferential membrane force must be the same. The membrane forces are equal to each other for every point of the shell:

$$n_{\phi\phi} = n_{\theta\theta} = \frac{1}{2}pa \tag{15.25}$$

The uniform normal load in the positive direction leads to tensile membrane forces, of which $n_{\phi\phi}$ works on the clamped edges, see Fig. 15.10b. We could have made a short cut for the computation of the membrane forces. The sphere cap can be considered as a part of a complete sphere with an internal overpressure p. This makes clear that $n_{\phi\phi}$ and $n_{\theta\theta}$ are equal. From the equilibrium Eq. (14.13) we had immediately obtained the result of Eq. (15.25). It also makes clear that the displacement u_{ϕ} is zero all over the sphere, and that u_z is nonzero and equal at all positions on the sphere surface. The value is easily obtained from the relation $u_z = \varepsilon_{\theta\theta}r$, where $r = a$. The constitutive relations in Eq. (14.21) yields the strain $\varepsilon_{\theta\theta}$:

$$\varepsilon_{\theta\theta} = \frac{(1-v)}{Et}n_{\theta\theta} = \frac{pa(1-v)}{2Et} \tag{15.26}$$

The normal displacement becomes

$$u_z = \varepsilon_{\theta\theta}a = \frac{(1-v)}{Et}n_{\theta\theta}a = \frac{pa^2(1-v)}{2Et} \tag{15.27}$$

The rotation φ_{ϕ} is calculated with Eq. (14.23):

$$\varphi_{\phi,m} = \frac{1}{r_1}\left(u_{\phi} - \frac{du_z}{d\phi}\right) = 0 \tag{15.28}$$

This is an evident result since the sphere cap is a part of the full sphere under overpressure. Finally, the horizontal displacement u_r of the edge, due to the membrane action, becomes with Eq. (15.17)

$$u_{r,m} = u_z \sin\phi_o = \frac{pa^2(1-v)}{2Et}\sin\phi_o \tag{15.29}$$

15.4.2 Bending Solution

The rotation of Eq. (15.28) and displacement of Eq. (15.29) of the membrane solution do not satisfy the boundary condition at the clamped edge, so an edge load f_h and an edge torque t_x are needed, which cause rotation and displacement equal to the membrane solution, however with opposite sign. The deformed shape of the sphere cap is as depicted in Fig. 15.10c. We assume positive directions as shown in Fig. 15.10d. This corresponds with the cone of Fig. 15.9b, therefore Eq. (15.21) applies:

$$\frac{1}{D_b}\begin{bmatrix} \dfrac{\sin^2\phi_o}{2\beta^3} & -\dfrac{\sin\phi_o}{2\beta^2} \\[2mm] -\dfrac{\sin\phi_o}{2\beta^2} & \dfrac{1}{\beta} \end{bmatrix}\begin{bmatrix} f_r \\ t_x \end{bmatrix} = \begin{bmatrix} u_r \\ \varphi_x \end{bmatrix}^b \qquad (15.30)$$

We must note that the $\phi\theta z$-coordinate system is adopted for the membrane analysis, and the xyz-system for bending. Therefore, an index x has the same meaning as an index ϕ; they both refer to the meridional direction, and accordingly the indices θ and y both refer to the circumferential direction. The total deformation at the clamped edge must be zero, thus:

$$u_{r(\phi=\phi_o)} = u_{r,m} + u_{r,b} = 0$$
$$\varphi_{\phi(\phi=\phi_o)} = \varphi_{\phi,m} + \varphi_{x,b} = 0 \qquad (15.31)$$

Hence, accounting for the Eqs. (15.29) and (15.28),

$$u_{r,b} = -u_{r,m} = -\frac{pa^2(1-\nu)}{2Et}\sin\phi_o \qquad (15.32)$$
$$\varphi_{x,b} = -\varphi_{\phi,m} = 0$$

If we introduce these values in Eq. (15.30), we can solve the edge force and torque:

$$f_r = -\frac{1-\nu}{2\beta\sin\phi_0}p$$
$$t_x = -\frac{1-\nu}{4\beta^2}p \qquad (15.33)$$

The negative signs indicate that both the force and the torque work in the opposite directions to the start directions in Fig. 15.10d. In order to calculate the forces and moments in the shell we must know the constant C and the phase angle ψ in Eqs. (15.5) and (15.6). For that purpose we must calculate f_z from f_r, see Fig. 15.8 and Eq. (15.17).

$$f_z = f_r\sin\phi_o = -\frac{1-\nu}{2\beta}p$$
$$t_x = -\frac{1-\nu}{4\beta^2}p \qquad (15.34)$$

Now we are able to calculate the constant C and the phase angle ψ. We know the value of two quantities, the rotation and the transverse shear force. On the basis of Eq. (15.6) the zero rotation yields $\psi = \pi/4$. The shear force v_x at the edge equals the edge force f_z; for $x = 0$ and $\psi = \pi/4$ we obtain from Eqs. (15.5) and (15.34) the condition

$$-\frac{1-v}{2\beta}p = -D\,2\sqrt{2}\beta^3 C. \tag{15.35}$$

The constant C is

$$C = \frac{1-v}{4\beta^4 D\,\sqrt{2}}p = \frac{1-v}{\sqrt{2}}\frac{pa^2}{Et} \tag{15.36}$$

Equation (15.5) become

$$m_{xx} = -\frac{1-v}{2\sqrt{2}}\frac{p}{\beta^2}e^{-\beta x}\sin(\beta x - \pi/4)$$

$$v_x = \frac{1-v}{2}\frac{p}{\beta}e^{-\beta x}\sin(\beta x - \pi/2) \tag{15.37}$$

$$n_{\theta\theta} = \frac{1-v}{\sqrt{2}}pa\,e^{-\beta x}\sin(\beta x + \pi/4)$$

For the complete solution we must add the membrane solution. This leads to

$$m_{xx} = -\frac{1-v}{2\sqrt{2}}\frac{p}{\beta^2}e^{-\beta x}\sin(\beta x - \pi/4)$$

$$v_x = \frac{1-v}{2}\frac{p}{\beta}e^{-\beta x}\sin(\beta x - \pi/2) \tag{15.38}$$

$$n_{\theta\theta} = pa\left(\frac{1}{2} + \frac{1-v}{\sqrt{2}}\,e^{-\beta x}\sin(\beta x + \pi/4)\right)$$

For a sphere with a thickness to radius ratio of $t/a = 1/30$ and an opening angle $\phi_o = 35°$, the meridional bending moment m_{xx} and the circumferential membrane force $n_{\theta\theta}$ are calculated with the aid of Eq. (15.38). The ratio of Poisson is assumed to be $v = 1/6$ for these calculations. Figures 15.11 and 15.12 show the result; the meridional bending moment in Fig. 15.11, and the membrane circumferential force (membrane hoop force) and circumferential force due to bending (denoted by hoop force due to bending) in Fig. 15.12. The membrane circumferential force is displayed with a negative sign, thus the actual circumferential force is the difference of the two lines. In the figures, the exact solution and the approximated solution by the edge disturbance bending theory are displayed. The exact solution is borrowed from Timoshenko's book "Theory of plates and shells" [4], and was obtained by not simplifying the equation for the homogeneous solution. The approximation is the solution of Eq. (15.38).

Since the thickness to radius ratio is large and the opening angle is comparatively small it can be concluded that the first approximation is accurate enough for most cases of structural interest. Also, the edge disturbance has practically vanished for this rather small value of β in the centre of this sphere cap. In a shell of revolution this is the opposite edge. This means that the edge disturbances starting at the edges do not influence each other in a considerable way and that the simplifications that are made to obtain the approximation are legitimate.

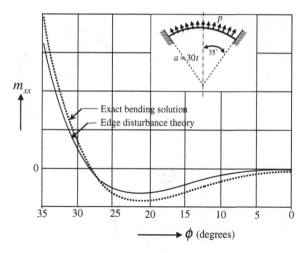

Fig. 15.11 Comparison of meridional bending moment of edge disturbance theory with exact solution

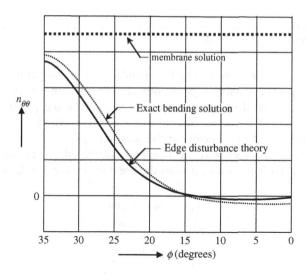

Fig. 15.12 Comparison of circumferential force of edge disturbance theory with exact bending solution

15.5   Application to a Pressured Hemispherical Boiler Cap

Consider a cylindrical boiler closed by a hemispherical end, and subject to an internal pressure p (Fig. 15.13). The shells have the same radius of curvature and it is assumed that the thicknesses of the shells are the same. We borrow this example from Flügge [5], and calculate it with the theory of this book. We choose an x_s-axis

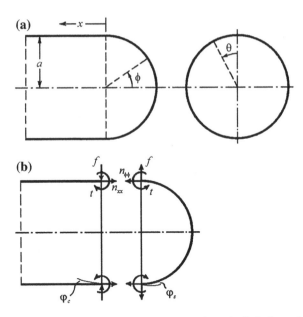

Fig. 15.13 Hemispherical boiler end. Axial and cross section (**a**). Cylinder and hemisphere cut apart to show relevant shell forces (**b**)

along the sphere and an x_c-axis along the cylinder, both starting at the junction, for the description of the bending edge disturbance. The membrane response of both shells leads to a normal displacement u_z of the shells at the junction, but these displacements do not have the same magnitude in the cylinder and the sphere. Since the edges have to fit together at the junction ($\phi = \phi_o$; $x_s = x_c = 0$), additional bending shell forces have to be applied to the edge. For the sphere, the membrane solution is the solution from the previous example with $\phi_o = \pi/2$:

$$n_{\phi\phi} = n_{\theta\theta} = \frac{1}{2}pa$$

$$\varphi_{\phi,m} = 0 \qquad\qquad (15.39)$$

$$u_{r,m} = \frac{pa^2(1-\nu)}{2Et}$$

For a cylinder, the circumferential membrane force is

$$n_{\theta\theta} = pam \qquad\qquad (15.40)$$

Though there is no axial load p_x, there is still a constant axial membrane force n_{xx}. Consider a cross-section over the cylinder perpendicular to its axis of revolution. The total force in the direction of this axis must be zero, because there is no external loading in this direction. The internal pressure delivers a resultant in this direction $p \cdot \pi a^2$ on the hemispherical end. Therefore, an equal tensile force is present in the cylindrical shell. This yields a membrane force

$$n_{xx} = \frac{p \cdot \pi a^2}{2\pi a} = \frac{1}{2}pa \tag{15.41}$$

The constitutive relation yields the circumferential strain $\varepsilon_{\theta\theta}$:

$$\varepsilon_{\theta\theta} = \frac{1}{Et}(-vn_{xx} + n_{\theta\theta}) = \frac{pa(1 - v/2)}{Et} \tag{15.42}$$

The displacement u_r in radial direction of the edge $(x = 0)$, due to membrane action, is

$$u_{r,m} = \varepsilon_{\theta\theta}a = \frac{pa^2(1 - v/2)}{Et} \tag{15.43}$$

Because of the constant load p and radius of the cylinder, the rotation φ_x of the cylinder at the junction is zero,

$$\varphi_{x,m} = 0 \tag{15.44}$$

For the cylinder, the membrane solution is

$$n_{xx} = \frac{1}{2}pa, \quad n_{\theta\theta} = pa$$
$$\varphi_{x,m} = 0, \quad u_{r,m} = \frac{pa^2(1 - v/2)}{Et} \tag{15.45}$$

By comparing the membrane solution Eq. (15.39) for the sphere with the membrane solution Eq. (15.45) for the cylinder, we see that the meridional membrane force and the rotation of the sphere and the cylinder appear to be equal to each other but the circumferential membrane forces are not. Consequently, the horizontal displacements are unequal. To obtain equal displacements and rotations at the junction, an edge force and torque are needed. These edge loads, which are shown in Fig. 15.13b denoted by f and t, can be interpreted as statically indeterminate loads. To obtain equilibrium, they are taken in the positive direction (see Fig. 15.13b) of their corresponding shell forces v_x and m_{xx}. The expressions for the radial displacement and the rotation of the sphere, due to the additional bending actions, are based on Eq. (15.21). With the angle $\phi_o = \pi/2$ we obtain for the edge of the sphere (subscript s)

$$\frac{1}{D_b}\begin{bmatrix} \dfrac{1}{2\beta_s^3} & -\dfrac{1}{2\beta_s^2} \\[2mm] -\dfrac{1}{2\beta_s^2} & \dfrac{1}{\beta_s} \end{bmatrix}\begin{bmatrix} f \\ t \end{bmatrix} = \begin{bmatrix} u_{r,b} \\ \varphi_{x,b} \end{bmatrix}_s \tag{15.46}$$

For the cylinder, the expressions for these displacements are based on Eq. (15.7), taking into account that $f_r = -f$ and that $t_x = -t$ (subscript c),

$$\frac{1}{D_b}\begin{bmatrix} \dfrac{1}{2\beta_c^3} & \dfrac{1}{2\beta_c^2} \\[2ex] \dfrac{1}{2\beta_c^2} & \dfrac{1}{\beta_c} \end{bmatrix}\begin{bmatrix} -f \\[1ex] -t \end{bmatrix} = \begin{bmatrix} u_{r,b} \\[1ex] \varphi_{x,b} \end{bmatrix}_c . \tag{15.47}$$

The shells have the same radius of curvature, a and r_y respectively, and the same thickness t. Because of this, β_s and β_c are equal:

$$\left.\begin{aligned} 4\beta_s^4 &= \frac{Et}{D_b r_y^2} \\[1ex] 4\beta_c^4 &= \frac{Et}{D_b a^2} \end{aligned}\right\} \quad \Rightarrow \quad \beta_s = \beta_c \tag{15.48}$$

We skip the subscripts c and s in β and obtain

$$\frac{1}{D_b}\begin{bmatrix} \dfrac{1}{2\beta^3} & -\dfrac{1}{2\beta^2} \\[2ex] -\dfrac{1}{2\beta^2} & \dfrac{1}{\beta} \end{bmatrix}\begin{bmatrix} f \\[1ex] t \end{bmatrix} = \begin{bmatrix} u_{r,b} \\[1ex] \varphi_{x,b} \end{bmatrix}_s \tag{15.49}$$

$$\frac{1}{D_b}\begin{bmatrix} -\dfrac{1}{2\beta^3} & -\dfrac{1}{2\beta^2} \\[2ex] \dfrac{1}{2\beta^2} & -\dfrac{1}{\beta} \end{bmatrix}\begin{bmatrix} f \\[1ex] t \end{bmatrix} = \begin{bmatrix} u_{r,b} \\[1ex] \varphi_{x,b} \end{bmatrix}_c \tag{15.50}$$

The deformation of the sphere and cylinder must be equal:

$$\begin{aligned} \left(u_{r,m} + u_{r,b}\right)_s &= \left(u_{r,m} + u_{r,b}\right)_c \\[1ex] \left(\varphi_{\phi,m} + \varphi_{x,b}\right)_s &= \left(\varphi_{x,m} + \varphi_{x,b}\right)_c \end{aligned} \tag{15.51}$$

With Eqs. (15.39) and (15.49) for the sphere and Eqs. (15.45) and (15.50) for the cylinder, these two equations become

$$\begin{aligned} &\frac{pa^2(1-v)}{2Et}\begin{bmatrix} 1 \\[1ex] 0 \end{bmatrix} + \frac{1}{D_b}\begin{bmatrix} \dfrac{1}{2\beta^3} & -\dfrac{1}{2\beta^2} \\[2ex] -\dfrac{1}{2\beta^2} & \dfrac{1}{\beta} \end{bmatrix}\begin{bmatrix} f \\[1ex] t \end{bmatrix} \\[3ex] &= \frac{pa^2(1-v/2)}{Et}\begin{bmatrix} 1 \\[1ex] 0 \end{bmatrix} + \frac{1}{D_b}\begin{bmatrix} -\dfrac{1}{2\beta^3} & -\dfrac{1}{2\beta^2} \\[2ex] -\dfrac{1}{2\beta^2} & -\dfrac{1}{\beta} \end{bmatrix}\begin{bmatrix} f \\[1ex] t \end{bmatrix} \end{aligned} \tag{15.52}$$

Rewriting these equations leads to

$$\frac{pa^2(1-v)}{2Et} + \frac{1}{2D_b\beta^2}\left(\frac{1}{\beta}f - t\right) = \frac{pa^2(1-v/2)}{Et} - \frac{1}{2D_b\beta^2}\left(\frac{1}{\beta}f + t\right)$$
$$-\frac{1}{D_b\beta}\left(\frac{1}{2\beta}f - t\right) = -\frac{1}{D_b\beta}\left(\frac{1}{2\beta}f + t\right)$$

(15.53)

And by rearranging we obtain

$$\frac{1}{D_b\beta^3}f = \frac{pa^2}{2Et}, \qquad \frac{1}{D_b\beta}t = 0$$

(15.54)

The two edge loads are

$$f = \frac{D_b\beta^3 a^2}{2Et}p = \frac{1}{8\beta}p, \qquad t = 0$$

(15.55)

The positive sign of the radial load indicates that this load works in the direction of the (positive) direction that is shown in Fig. 15.13. This is to be expected since the membrane displacement of the cylinder is larger than the membrane displacement of the sphere. Furthermore, neither shell undergoes any rotation due to the membrane action. The bending moment at the junction is zero because the horizontal loads close the 'membrane gap' and lead to equal rotation of the sphere and the cylinder, in other words to the same slope of the meridians. This feature occurs because of the fact that the shells have the same radius of curvature and thickness.

For the meridional bending moment m_{xx} and the transverse shear force v_x of the sphere, Eq. (15.37) is used. With $x_s = 0$ at the junction, these expressions, accounting for Eq. (15.55), yield respectively

$$t = m_{xx} = -2D_b\beta^2 C_s \sin\left(\psi_s - \frac{\pi}{2}\right) = 0$$
$$f = v_x = -2\sqrt{2}D_b\beta^3 C_s \sin\left(\psi_s - \frac{3\pi}{4}\right) = \frac{1}{8\beta}p$$

(15.56)

The phase angle ψ_s and the constant C_s are

$$\psi_s = \frac{\pi}{2}, \qquad C_s = \frac{p}{16D_b\beta^4} = \frac{pa^2}{4Et}$$

(15.57)

With this solution, the radial displacement u_r and the rotation φ_x at the edge of the sphere are derived with Eq. (15.6):

$$u_r = u_z = C_s \sin\psi_s = \frac{pa^2}{4Et}$$
$$\varphi_x = -\beta\sqrt{2}C_s \sin\left(\psi_s - \frac{\pi}{4}\right) = -\frac{pa^2\beta}{4Et}$$

(15.58)

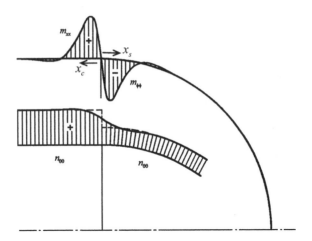

Fig. 15.14 Bending moments m_{xx} and membrane force $n_{\theta\theta}$ at the junction

For the meridional bending moment m_{xx} and the transverse shear force v_x of the cylinder, Eq. (15.5) is used. With $x_c = 0$ at the junction, the expressions, accounting for Eq. (15.55), yield respectively:

$$t = m_{xx} = -2D_b\beta^2 C_c \sin\left(\psi_c - \frac{\pi}{2}\right) = 0$$

$$f = v_x = 2\sqrt{2}D_b\beta^3 C_c \sin\left(\psi_c - \frac{3\pi}{4}\right) = \frac{1}{8\beta}p \tag{15.59}$$

The phase angle ψ_c and the constant C_c are thus equal to:

$$\psi_c = \frac{\pi}{2}; \qquad C_c = -\frac{p}{16D_b\beta^4} = -\frac{pa^2}{4Et} \tag{15.60}$$

With this solution, the radial displacement u_r and the rotation φ_x at the edge of the sphere are derived with Eq. (15.6):

$$u_r = C_c \sin\psi_c = -\frac{pa^2}{4Et}$$

$$\varphi_x = \beta\sqrt{2}C_c \sin\left(\psi_c - \frac{\pi}{4}\right) = -\frac{pa^2\beta}{4Et} \tag{15.61}$$

By using membrane solutions Eqs. (15.39) and (15.45) and solution Eqs. (15.57) and (15.60) for the constants of the edge disturbance, we can derive the meridional bending moment and the circumferential membrane force in the sphere and in the cylinder. For the sphere, we obtain

$$n_{\theta\theta} = pa\left(\frac{1}{2} + \frac{1}{4}e^{-\beta x_s} \cos\beta x_s\right)$$

$$m_{\phi\phi} = -\frac{p}{8\beta^2} e^{-\beta x_s} \sin\beta x_s \tag{15.62}$$

For the cylinder, we obtain

$$n_{\theta\theta} = pa\left(1 - \frac{1}{4}e^{-\beta x_c}\cos\beta x_c\right)$$

$$m_{xx} = \frac{p}{8\beta^2}e^{-\beta x_c}\sin\beta x_c$$

(15.63)

These shell forces are shown in Fig. 15.14 for Poisson's ratio $v = 1/3$ and $t/a = 1/100$. The membrane theory yields a discontinuity in the $n_{\theta\theta}$-diagram (the broken line). The bending action of the shells removes this discontinuity and leads to a continuous transition. The edge load f, needed to close the 'membrane gap', leads to bending moments in the shells. The bending moment is zero at the boundary between the cylinder and the sphere. At a closer look, the considered problem is an elementary case b of Fig. 5.12. There we see that the maximum bending moment occurs at a distance $x = \pi/(4\beta)$, and the maximum value is, accounting for Eq. (15.55),

$$m_{\phi\phi,\max} = 0.322f/\beta = 0.0403p/\beta^2$$

(15.64)

At the same place the value of the membrane force due to bending is calculated:

$$n_{\theta\theta} = 0.0285\ pa.$$

(15.65)

The maximum stress in the shell becomes (we take $v = 0.3$)

$$\sigma_{\phi\phi} = \frac{n_{\phi\phi}}{t} + \frac{6m_{\phi\phi}}{t^2} = 0.5\frac{ap}{t} + 6\times0.0403\frac{p}{\beta^2t^2}$$

$$= \left(0.5 + \frac{6\times0.0403}{\sqrt{3(1-v^2)}}\right)\frac{ap}{t} = (0.5+0.1463)\frac{ap}{t} = 0.646\frac{ap}{t}$$

(15.66)

The stress due to the bending stress is 29 % of the membrane stress $0.500\ pa/t$ outside the bending edge zone. For the cylinder we obtain

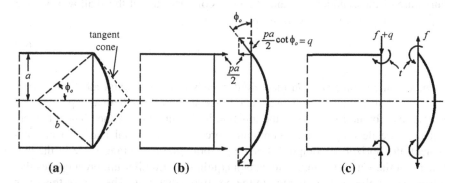

Fig. 15.15 Shallow spherical boiler end, axial section (**a**), the membrane forces in the principal system (**b**) and edge loads of the additional bending action (**c**)

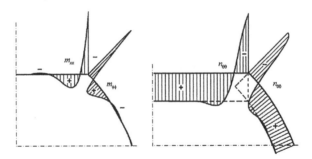

Fig. 15.16 Shell forces $m_{\phi\phi}(m_{xx})$ and $n_{\theta\theta}$ at the junction

$$
\begin{aligned}
\sigma_{\phi\phi} &= \frac{n_{\phi\phi}}{t} + \frac{6m_{\phi\phi}}{t^2} = \frac{ap}{t} + 6 \times 0.0403 \frac{p}{\beta^2 t^2} \\
&= \left(1 + \frac{6 \times 0.0403}{\sqrt{3(1 - v^2)}}\right) \frac{ap}{t} = (1 + 0.146) \frac{ap}{t} = 1.146 \frac{ap}{t}
\end{aligned}
\tag{15.67}
$$

This value is 15 % higher than the membrane stress outside the bending edge zone.

15.6 Application to a Pressured Shallow Spherical Boiler Cap

Consider a cylindrical boiler closed by a shallow spherical end, subject to an internal pressure p (Fig. 15.15). The opening angle of the cap is ϕ_o. The spherical cap meets the cylinder under an angle. In Fig. 15.16 the conical shell, which is the tangent plane at the junction, is shown with dashed lines. The thickness of the cylindrical shell is equal to the thickness of the spherical shell. The radius of curvature of the cylinder is a, and the radius of curvature of the shallow spherical cap is b. The relation between the two radii is

$$
b = a / \sin \phi_o
\tag{15.68}
$$

A pure membrane state in the cylinder leads to only an axial membrane force $n_{xx} = pa/2$ at the junction. A pure membrane state in the spherical cap leads to a normal membrane force $n_{\phi\phi}$ at the junction in the direction of the conical tangent plane. If we decompose this membrane force in a horizontal and vertical component, the horizontal component can equilibrate the membrane force in the cylinder, but the vertical component cannot equilibrate. Equilibrium presupposes that a load q is applied at the junction in the positive z-direction. The size of this force is $q = (pa/2) \cot \phi_o$. In reality this force is not there, so we must correct the load state a load q in the negative direction:

$$q = -\frac{pa}{2} \cot \phi_o \tag{15.69}$$

We choose to apply this load to the edge of the cylinder (refer to Fig. 15.15c). It causes a radial displacement $u_{r,q}$ and a rotation $\varphi_{x,q}$ of the cylinder edge. To calculate these values, we use the second row of elementary case b of Table 5.1:

$$\frac{1}{2D_b\beta^2} \begin{bmatrix} -1/\beta \\ -1 \end{bmatrix} q = \begin{bmatrix} u_{r,q} \\ \varphi_{x,q} \end{bmatrix} \tag{15.70}$$

The membrane response of both shells, in combination with the applied load q, leads to a horizontal displacement of the shells at the junction, but these displacements do not have the same magnitude. Since the edges have to fit together at the junction ($\phi = \phi_o$; $x = 0$), additional bending stress resultants have to be applied to the edge.

The membrane solution for a sphere (with radius of curvature b) submitted to an internal pressure is the membrane solution from Sect. 15.5 and it reads:

$$n_{\phi\phi} = n_{\theta\theta} = \frac{1}{2}pb$$

$$\varphi_{\phi,m} = 0 \tag{15.71}$$

$$u_{r,m} = \frac{pb^2(1-\nu)}{2Et} \sin \phi_o$$

The membrane solution for a cylinder (with radius of curvature a) subject to an internal pressure is the membrane solution from Sect. 4.9:

$$n_{xx} = \frac{1}{2}pa, \quad n_{\theta\theta} = pa$$

$$\varphi_{x,m} = 0, \quad u_{r,m} = \frac{pa^2(1-\nu/2)}{Et} \tag{15.72}$$

The load q that is applied to the edge of the cylinder to procure the equilibrium leads to a horizontal displacement and a rotation of the edge of the cylinder of Eq. (15.70):

$$\frac{1}{2D_b\beta^2} \begin{bmatrix} -\frac{1}{\beta} \\ -1 \end{bmatrix} q = \begin{bmatrix} u_{r,q} \\ \varphi_{x,q} \end{bmatrix}_c \tag{15.73}$$

To obtain equal displacement and rotation of the edges of the sphere and cylinder, an edge load f and torque t are needed. These are chosen as shown in Fig. 15.15c, corresponding with the shell forces v_x and m_{xx}. These forces lead to edge displacements and rotations in the shell cap and the cylinder. In the shell cap we obtain

$$\frac{1}{D_b}\begin{bmatrix} \dfrac{1}{2\beta_s^3}\sin^2\phi_o & -\dfrac{1}{2\beta_s^2}\sin\phi_o \\[2mm] -\dfrac{1}{2\beta_s^2}\sin\phi_o & \dfrac{1}{\beta_s} \end{bmatrix}\begin{bmatrix} f \\[2mm] t \end{bmatrix} = \begin{bmatrix} u_{r,b} \\[2mm] \varphi_{x,b} \end{bmatrix}_s \tag{15.74}$$

At the edge of the cylinder we obtain

$$\frac{1}{D_b}\begin{bmatrix} \dfrac{1}{2\beta_c^3} & \dfrac{1}{2\beta_c^2} \\[2mm] \dfrac{1}{2\beta_c^2} & \dfrac{1}{\beta_c} \end{bmatrix}\begin{bmatrix} -f \\[2mm] -t \end{bmatrix} = \begin{bmatrix} u_{r,b} \\[2mm] \varphi_{x,b} \end{bmatrix}_c \tag{15.75}$$

We can calculate f and t by the following condition:

$$\begin{aligned} \left(u_{r,m}+u_{r,b}\right)_s &= \left(u_{r,m}+u_{r,b}\right)_c+\left(u_{r,q}\right)_c \\ \left(\varphi_{\phi,m}+\varphi_{x,b}\right)_s &= \left(\varphi_{x,m}+\varphi_{x,b}\right)_c+\left(\varphi_{x,q}\right)_c \end{aligned} \tag{15.76}$$

If we substitute the Eqs. (15.71) up to and including (15.75), we obtain two equations with the values f and t:

$$\begin{aligned} f &= -\frac{1}{\left(1+\sqrt{\sin\phi_o}\right)}\frac{pa}{2}\cot\phi_o \\[2mm] t &= -\frac{\sin\phi_o}{\left(1+\sqrt{\sin\phi_o}\right)}\frac{pa}{4\beta_s}\cot\phi_o \end{aligned} \tag{15.77}$$

To obtain this solution, a small term has been neglected, which limits the solution to opening angles $(\pi/8) < \phi_0 < (3\pi/8)$. The negative sign of the edge force and torque indicates that these shell forces work in the opposite directions of the (positive) directions that are shown in Fig. 15.15c. The transverse shear force v_x and the bending moment m_{xx} at the junction become

$$\begin{aligned} v_{x(x=0)} &= q+f = \frac{\sqrt{\sin\phi_0}}{\left(1+\sqrt{\sin\phi_0}\right)}\frac{pa}{2}\cot\phi_0 \\[2mm] m_{xx(x=0)} &= t = -\frac{\sin\phi_0}{\left(1+\sqrt{\sin\phi_0}\right)}\frac{pa}{4\mu}\cot\phi_0 \end{aligned} \tag{15.78}$$

In Fig. 15.16 the meridional bending moment is shown for Poisson's ratio $v = 1/3$ and a thickness to radius ratio 1/100 (as used in the previous example). For the opening angle of the sphere cap, we have taken $\phi_o = \pi/4$. The $n_{\theta\theta}$-diagram shows that in the zone of the edge disturbance high compressive stresses are developed on both sides of the edge. Compared to the previous example, the distribution of the bending stress resultants is also entirely different. The bending moment is not zero at the junction between the shells, but has a sharp peak there and the scale of this figure is therefore not the same as the scale of the previous example. Compared to the previous example, the bending stresses are far more important now. The ratio between the bending moments is

$$\frac{m_{\max(shallow\ sphere)}}{m_{\max(hemisphere)}} \simeq 36. \tag{15.79}$$

It is clearly not advisable to have a sharp edge between the boiler end and the boiler drum. An almost sharp edge in the meridian, rounded by an arc of great curvature, has little advantage. To reduce the bending stresses in the shell, a stiffening ring must at least be provided at the edge.

References

1. Geckeler JW (1930) Zur Theorie der Elastizität flacher rotationssymmetrischer Schalen. Ing-Arch 1:255–270
2. Geckeler JW (1926) Ueber die Festigkeit achsensymmetrischer Schalen, Forschungsarbeiten. Ingenieurswesen, Heft 276, Berlin (in German)
3. Pucher A (1937/1938) Die Berechnung der Dehnungspannungen von Rotationsschalen met Hilfe von Spannungsfunktionen, Mémoires AIPC V
4. Timoshenko SP, Woinowsky-Krieger S (1959) Theory of plates and shells, 2nd edn. Mc Graw-Hill Book Company Inc., New York
5. Flügge W (1960) Stresses in shells. Springer, Berlin (corrected reprint 1962)

Part V
Capita Selecta

Chapter 16
Introduction to Buckling

16.1 Problem Statement

The investigation of stability of shell structures is a specialty which in its own right deserves a complete separate book. Here just some main aspects are explained, and the correspondence and difference with beam-column buckling is touched. We will successively discuss buckling of uni-axially loaded plates as a limit case of shells, arched beams, arched circular roofs, axially-pressed shells of rotation and domes.

16.2 Beam-Column Buckling

In order to quickly understand how the Donnell equation for shells must be adapted, we call to memory how the differential equation for beam-columns is obtained. Consider the elementary beam-column element of length dx shown in Fig. 16.1. It is loaded by a lateral distributed load q and normal compressive forces N at each end of the element, and we neglect shear deformation. The differential equation of the beam-column is

$$EI\frac{d^4u_z}{dx^4} = q - q_N \qquad (16.1)$$

Herein EI is the bending rigidity, u_z the lateral displacement and q_N the effect of the normal force N. If the normal force is zero, q_N vanishes. For a nonzero normal force and positive second derivative of the deflection line, q_N is positive. The value of q_N follows from the fact that the normal forces N at the ends of the element make different angles with the beam-column axis in the unloaded state. This results in a distributed load per unit length of the size, see Fig. 16.1,

J. Blaauwendraad and J. H. Hoefakker, *Structural Shell Analysis*,
Solid Mechanics and Its Applications 200, DOI: 10.1007/978-94-007-6701-0_16,
© Springer Science+Business Media Dordrecht 2014

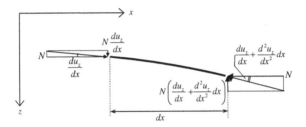

Fig. 16.1 A negative distributed load $N\, d^2u_z/dx^2$ results at a positive curvature

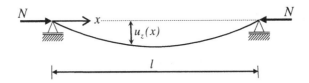

Fig. 16.2 Buckled simply-supported beam-column

$$q_N = N\frac{d^2u_z}{dx^2} \tag{16.2}$$

Substitution of Eq. (16.2) into Eq. (14.1) and reordering yields

$$EI\frac{d^4u_z}{dx^4} + N\frac{d^2u_z}{dx^2} = q \tag{16.3}$$

For a simply-supported beam-column of span l and the origin of the x-axis at the middle of the span, shown in Fig. 16.2, we adopt the deflection shape

$$u_z = \hat{u}_z \cos\frac{\pi x}{l} \tag{16.4}$$

Substitution of this deflection in Eq. (16.3) yields for zero distributed load q the eigenvalue problem

$$\left(EI\frac{\pi^4}{l^4} - N\frac{\pi^2}{l^2}\right)u_z = 0 \tag{16.5}$$

from which we obtain the well-known critical Euler buckling load

$$N_{cr} = \frac{\pi^2 EI}{l^2} \tag{16.6}$$

16.3 Shell Buckling Study on the Basis of Donnell Equation

We derived the Donnell equation for shallow shells in Chap. 6. In Eq. (6.20) we ended up with two coupled differential equations, which are for zero distributed loads p_x and p_y

$$-\Gamma\Phi + D_b\Delta^2\Delta^2 u_z = p_z \tag{16.7}$$

$$\Delta^2\Delta^2\Phi + D_m(1 - v^2)\Gamma u_z = 0 \tag{16.8}$$

where

$$\Gamma = k_x\frac{\partial^2}{\partial y^2} - 2k_{xy}\frac{\partial^2}{\partial x\partial y} + k_y\frac{\partial^2}{\partial x^2}$$
$$\Delta^2 = \frac{\partial^2}{\partial x^2} + \frac{\partial^2}{\partial y^2} \tag{16.9}$$

Equation (16.7) describes the equilibrium in the direction normal to the shell surface. For zero shell curvatures k_x, k_y and k_{xy} the operator Γ vanishes, and the equation reduces to the biharmonic equation of thin flexural plates. To study stability we add contributions due to the membrane forces n_{xx}, n_{yy} and n_{xy} to Eq. (16.7), comparable to what we did in Eq. (16.3) for beam-columns. At the same time we put q zero. We obtain

$$-\Gamma\Phi + D_b\Delta^2\Delta^2 u_z + n_{xx}\frac{\partial^2 u_z}{\partial x^2} + 2n_{xy}\frac{\partial^2 u_z}{\partial x\partial y} + n_{yy}\frac{\partial^2 u_z}{\partial y^2} = 0 \tag{16.10}$$

As we did in Sect. 6.5.3, we eliminate Φ from Eqs. (16.8) and (16.10):

$$\frac{Et^3}{12}\Delta^2\Delta^2\Delta^2\Delta^2 u_z + Et\,\Gamma^2 u_z + \Delta^2\Delta^2\left(n_{xx}\frac{\partial^2 u_z}{\partial x^2} + 2n_{xy}\frac{\partial^2 u_z}{\partial x\partial y} + n_{yy}\frac{\partial^2 u_z}{\partial y^2}\right) = 0 \tag{16.11}$$

In this equation, we substituted values D_b and D_b for zero Poisson's ratio. More refinement is not needed for the purpose of this chapter, considering other assumptions in the derivation. Equation (16.11) is the starting point of our stability investigations. The forces n_{xx}, n_{yy} and n_{xy} can be calculated with the linear theory. It holds in general that the buckled area in shells often is not close to the edges; it is small compared with the dimensions of the total shell.

Furthermore it is observed that the direction of the axes of the buckled area is in the direction of the axes of the stresses and that no big error is made if we neglect the shear force n_{xy}.

For the calculation of the buckling load, we introduce an area with length a and width b. We choose the origin of the x- and y-axis in the centre of the area. We introduce the buckled deflection shape

$$u_z = \hat{u}_z \cos \frac{\pi x}{a} \cos \frac{\pi y}{b} \qquad (16.12)$$

Herein a is the wavelength in x-direction and b in y-direction. Substitution of this shape in Eq. (16.11) yields an eigenvalue problem from which we obtain, accounting for $k_x = r_x^{-1}$ and $k_y = r_y^{-1}$,

$$\left(\frac{\pi^2}{a^2} + \frac{\pi^2}{b^2}\right)^2 \left(\frac{\pi^2}{a^2} n_{xx} + \frac{\pi^2}{b^2} n_{yy}\right) = \frac{Et^3}{12} \left(\frac{\pi^2}{a^2} + \frac{\pi^2}{b^2}\right)^4 + Et \left(\frac{\pi^2}{a^2 r_y} + \frac{\pi^2}{b^2 r_x}\right)^2 \qquad (16.13)$$

Herein we introduced the radii r_x and r_y which are the reciprocals of k_x and k_y. Equation (16.13) will be our basis for the calculation of the critical buckling load. From this we can derive some simple cases by reduction.

16.4 Buckling Check for Beam-Column

Consider again the simply-supported beam-column of span l in Fig. 16.2. We can simulate the buckling of this structure by making r_x, r_y and the span b in y-direction infinitely large:

$$r_x = \infty; \quad r_y = \infty; \quad b = \infty; \quad EI = \frac{Et^3}{12}; \quad a = l \qquad (16.14)$$

This reduces Eq. (16.13) to

$$\left(\frac{\pi^2}{a^2}\right)^2 \left(\frac{\pi^2}{a^2} n_{xx}\right) = \frac{Et^3}{12} \left(\frac{\pi^2}{a^2}\right)^4 \qquad (16.15)$$

from which we obtain, replacing n_{xx} by N,

$$N_{cr} = \frac{\pi^2 EI}{l^2} \qquad (16.16)$$

We conclude that Eq. (16.13) reduces correctly to the Euler buckling load for beam-columns.

16.5 Check for Flat Plate

Consider the long plate width span l shown in Fig. 16.3 which is simply-supported along two parallel edges and axially loaded by a membrane force n_{xx}. For this case it holds that

$$n_{yy} = 0; \quad r_x = r_y = 0 \qquad (16.17)$$

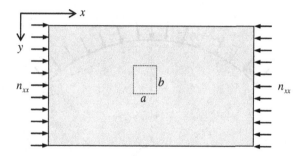

Fig. 16.3 Axially-pressed flat plate

which reduces Eq. (14.15) to

$$\left(\frac{\pi^2}{a^2}+\frac{\pi^2}{b^2}\right)^2\left(\frac{\pi^2}{a^2}n_{xx}\right)=\frac{Et^3}{12}\left(\frac{\pi^2}{a^2}+\frac{\pi^2}{b^2}\right)^4 \tag{16.18}$$

and we obtain

$$n_{xx,cr}=\frac{\pi^2 Et^3}{12\,b^2}\left(\frac{b}{a}+\frac{a}{b}\right)^2 \tag{16.19}$$

We recall that a is the wavelength in axial direction and b the wavelength in span direction. We want to know for which ratio b/a the critical buckling load is minimal. An expression $x+x^{-1}$ has a minimum value 2 for $x=1$. So we obtain the minimum for the ratio $a/b=1$:

$$n_{xx,cr}=\frac{\pi^2 Et^3}{3\,b^2} \tag{16.20}$$

The lowest value of this expression is reached when the wavelength b is equal to the span l in the direction of b. Supposing that an equal wavelength a can develop in the other direction, the result is:

$$n_{xx,cr}=\frac{\pi^2 Et^3}{3\,l^2} \tag{16.21}$$

16.6 Arch Buckling

Consider the circular arch shown in Fig. 16.4 which is pin-connected to fixed supports. The x-axis is chosen along the arch. The radius is r_x and the opening angle ϕ_0. A constant membrane force n_{xx} is present in the arch. We feed Eq. (16.13) with

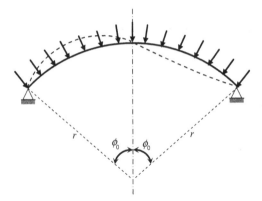

Fig. 16.4 Buckling mode of arch

$$r_y = \infty; \quad b = \infty; \quad n_{yy} = 0 \tag{16.22}$$

which reduces it to

$$\left(\frac{\pi^2}{a^2}\right)^2 \left(\frac{\pi^2}{a^2} n_{xx}\right) = \frac{Et^3}{12} \left(\frac{\pi^2}{a^2}\right)^4 \tag{16.23}$$

We obtain

$$n_{xx,cr} = \frac{\pi^2 Et^3}{12\, a^2} \tag{16.24}$$

The last step is to find the minimal possible value of a. The arch will not buckle over its full length, because this requires elongation of the arch, which is prevented by the large extensional rigidity. Buckling in two waves as shown in Fig. 16.4 is possible without elongation of the arch. Therefore we obtain $a = \phi_0\, r_x$. Further we introduce $EI = Et^3/12$, so the critical buckling load of the arch becomes, replacing n_{xx} by N,

$$N_{cr} = \frac{\pi^2 Et^3}{12\,(\phi_0\, r_x)^2} \tag{16.25}$$

16.7 Buckling of a Curved Plate and Cylinder Under Lateral Pressure

Consider the cylindrically curved plate loaded perpendicular to its surface by a homogenous pressure p shown in Fig. 16.5. The x-axis is chosen in the curved direction. The radius is r. Due to the load p, a membrane force $n_{xx} = r_x p$ will

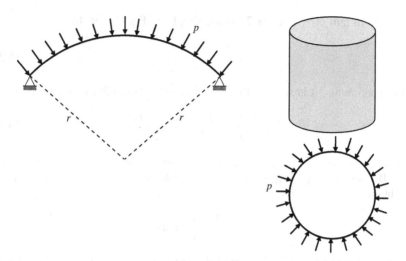

Fig. 16.5 Buckling of curved plate and cylinder under pressure

occur. We expect a buckling mode in which both a and b have finite values. We feed Eq. (16.13) with

$$r_x = r; \quad r_y = \infty; \quad n_{yy} = 0 \tag{16.26}$$

which changes it into

$$\left(\frac{\pi^2}{a^2} + \frac{\pi^2}{b^2}\right)^2 \left(\frac{\pi^2}{a^2} n_{xx}\right) = \frac{Et^3}{12}\left(\frac{\pi^2}{a^2} + \frac{\pi^2}{b^2}\right)^4 + Et\left(\frac{\pi^2}{b^2 r}\right)^2 \tag{16.27}$$

The critical pressure is $p_{cr} = n_{xx}/r$. From Eq. (16.27) we obtain

$$p_{cr} = \frac{n_{xx,cr}}{r} = \frac{Et^3}{12}\frac{a^2}{r\pi^2}\left(\frac{\pi^2}{a^2} + \frac{\pi^2}{b^2}\right)^2 + Et\frac{a^2}{r\pi^2}\left(\frac{\pi^2}{b^2 r}\right)^2 \left(\frac{\pi^2}{a^2} + \frac{\pi^2}{b^2}\right)^{-2} \tag{16.28}$$

We rewrite this expression into

$$p_{cr} = \frac{b^2}{a^2 r^2}\frac{Et^2}{\sqrt{12}}\left(A + A^{-1}\right) \tag{16.29}$$

where

$$A = \frac{\pi^2 rt}{\sqrt{12}\,b^2}\left(\frac{b}{a} + \frac{a}{b}\right)^2 \tag{16.30}$$

The minimum of $A + A^{-1}$ is 2, which simplifies Eq. (16.29) to

$$p_{cr} = \frac{b^2}{a^2 r^2} \frac{E t^2}{\sqrt{3}} \tag{16.31}$$

In experiments, it is seen that $a \approx b$. This further simplifies the expression to

$$p_{cr} = \frac{1}{\sqrt{3}} E \left(\frac{t}{r}\right)^2 \approx 0.6 E \left(\frac{t}{r}\right)^2 \tag{16.32}$$

We can modify Eq. (16.32) to a critical value for the stress σ_{xx} in the curved plate. Because of $\sigma_{xx} = n_{xx}/t$ and $n_{xx} = p r$, it holds that $\sigma_{xx} = p(r/t)$. Therefore we obtain

$$\sigma_{xx,cr} = \frac{1}{\sqrt{3}} E \frac{t}{r} \approx 0.6 E \frac{t}{r} \tag{16.33}$$

It is possible to calculate the size of the wavelength on the basis of Eq. (16.30). The minimum value of A is 1. So, for $a \approx b$ Eq. (16.30) becomes

$$\frac{2 \pi^2 r t}{\sqrt{3} b^2} = 1 \tag{16.34}$$

from which we obtain

$$b \approx 3.4 \sqrt{r t} \tag{16.35}$$

16.8 Buckling of Axially-Pressed Cylinder

Finally we consider the buckling of an axially pressed circular cylinder as shown in Fig. 16.6. We choose the y-axis in axial direction. The load results in a membrane force n_{xx}. We feed Eq. (16.13) with

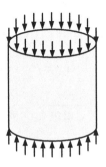

Fig. 16.6 Buckling of axially pressed cylinder

$$r_x = \infty; \quad r_y = r; \quad n_{yy} = 0 \tag{16.36}$$

which changes it into

$$\left(\frac{\pi^2}{a^2} + \frac{\pi^2}{b^2}\right)^2 \left(\frac{\pi^2}{a^2} n_{xx}\right) = \frac{Et^3}{12}\left(\frac{\pi^2}{a^2} + \frac{\pi^2}{b^2}\right)^4 + Et\left(\frac{\pi^2}{b^2 r}\right)^2 \tag{16.37}$$

We solve n_{xx}:

$$n_{xx} = \frac{Et^3}{12}\frac{a^2}{\pi^2}\left(\frac{\pi^2}{a^2} + \frac{\pi^2}{b^2}\right)^2 + Et\frac{a^2}{\pi^2}\left(\frac{\pi^2}{b^2 r}\right)^2 \left(\frac{\pi^2}{a^2} + \frac{\pi^2}{b^2}\right)^{-2} \tag{16.38}$$

This is the same expression as we have seen before in Eq. (16.28). There we have found the solution

$$\sigma_{xx,cr} = 0.6E\frac{t}{r} \tag{16.39}$$

16.9 Buckling of Spheres, Hyppars and Elpars Subject to Lateral Pressure

Zoelly [1] has shown in 1915 that Eq. (16.39) applies also for spheres under a pressure normal to the shell surface, see Fig. 16.7. The same formula even appears to apply to hyperbolic paraboloids (hyppars) and elliptic paraboloids (elpars).

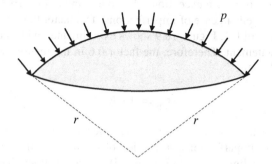

Fig. 16.7 Buckling of spherical shell

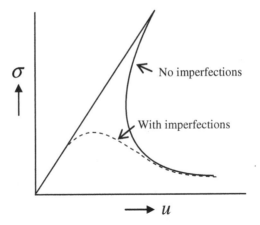

Fig. 16.8 Post-buckling behaviour of shell and effect of imperfection

16.10 Reducing Effect of Imperfections

Equation (16.39) can be considered as a general expression for the instability of shells, which even applies for double curved shells. We must stress an important point. There is a big difference with beam-column buckling in the post buckling behaviour. After a beam-column buckles, the critical buckling load stays sustained for substantial lateral buckling deflections. This is in general not the case for shells. Most shells show behaviour as depicted in Fig. 16.8 where the membrane stress is shown as a characteristic displacement, for instance the shortening of an axially pressed cylinder. After the critical stress is reached a sudden snap-back takes place and the post-buckling load carrying capacity collapses drastically to a much lower level. As a consequence, shells are very sensitive to imperfections. Only hyppars are a fortunate exception. The dotted line in Fig. 16.8 shows the real response of most shells in case of imperfections. The factor 0.6 may reduce to low values in the order of 0.1. Figure 16.9 shows experimental evidence of Zoelly [1], supporting the statement. Therefore, the factor 0.6 in the theoretical expression is better decreased to 0.1:

$$\sigma_{xx,cr} = 0.1\ E\frac{t}{r} \qquad\qquad (16.40)$$

In codes of practice, the factor may be made dependent on the shell type and the size of the inward imperfections. Inspection of the imperfections requires insight in the expected buckling waves. In Eq. (16.35) we have learned that the size of the wavelength is between 3 and 4 times $\sqrt{rt}$. This is useful information. Imperfections in cylinders must be measured with a straight rod or circularly curved template of specified length. The rod applies for meridians and the template for

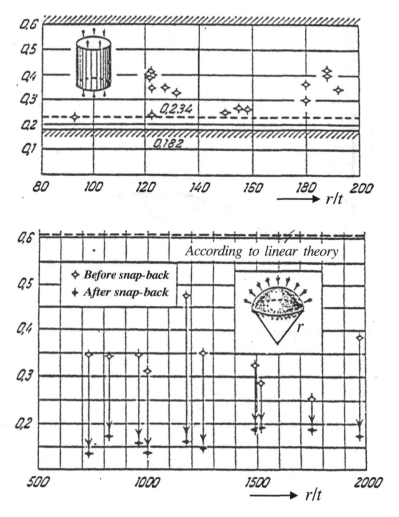

Fig. 16.9 Experimental evidence of effect of imperfection according to [1]. Notation and language adapted

parallel circles. Figure 16.10 shows an example as given in the European Recommendation for Steel Structures. The prescribed rod length is $4\sqrt{rt}$. Stress reduction factors may be limited depending on the ratio of the largest measured inward amplitude $\bar{u}$ and l_r.

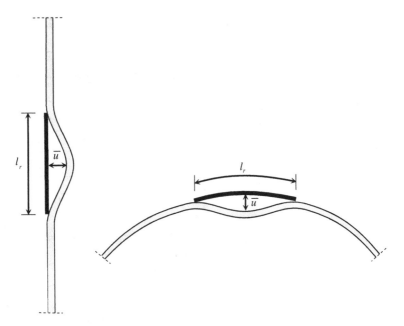

Fig. 16.10 Measurement of imperfections

Reference

1. Zoelly R (1915) Über ein Knickproblem an der Kugelschale, 1915. Dissertation, Zurich

Chapter 17
FEA for Shells of Irregular Shape

In the preceding chapters we have discussed analytical methods that can be used for shells with a shape that can be described by simple geometrical functions: spheres, cones, hyperbolic-paraboloids, cylinders, etc. Architects, however, do not feel restricted to these mathematical expressions. Modern shell shapes are freely chosen and may be more complicated. Figure 17.1 is an example, which shows a shell out of a row of shells. The shell has a rectangular ground plan and is simply supported along two sides. The other two sides are lines of symmetry in which the shell runs on continuously to its neighbouring shells. There, no supports occur. The circular top edge of the shell is a clamped support. For a shell of this complicated shape, a Finite Element Analysis (FEA) is the only way to compute the stress distribution. In Sect. 17.1 we will briefly discuss the principles of the Finite Element Method for a linear-elastic shell, and after that, demonstrate in Sect. 17.2 the calculation of the irregular shell in Fig. 17.1.

17.1 Finite Element Analysis

In a Finite Element Analysis we need not make distinctions between membrane solutions and bending solutions. The method always accounts for combined membrane and bending deformation and even may take account of transverse shear deformation. Shell finite elements do in general not distinguish between thin and thick elements, and the application is not restricted to shallow shell types. We will not discuss here the very internals of FEA, but restrict ourselves to main lines. The shell surface is discretized into a big number of finite elements. In general, elements are doubly curved and have triangular or quadrilateral shape as shown in Fig. 17.2. Nodes may occur in the corners and mid-side of the edges. Shells of any irregular shape can be modelled with aid of these elements. In this book, we consider static loads only. Loads may consist of distributed loads over the element area, line loads along element edges and point loads in nodes.

Finite Element Analysis is an approximation, with higher accuracy for larger numbers of elements. Tractions on the element edges are lumped to element forces

J. Blaauwendraad and J. H. Hoefakker, *Structural Shell Analysis*,
Solid Mechanics and Its Applications 200, DOI: 10.1007/978-94-007-6701-0_17,
© Springer Science+Business Media Dordrecht 2014

Fig. 17.1 Part out of a row of shells. The short edges are supported, the long edges are unsupported lines of symmetry

Fig. 17.2 Quadrilateral and triangular shell element

in the nodes. Distributed loading over the element area and line loads along element edges are lumped to nodal loads. In general nodes are connected to one or more adjacent elements. It holds, that the sum of the element forces of all elements meeting in a node and the applied load at that node are in equilibrium. On the interface of adjacent elements, full continuity of displacements occurs. The kinematic equations and the constitutive laws are always fully satisfied within elements, but the equilibrium equations, in general, are not. Element equilibrium is satisfied only in a weak way, such that the virtual work of the stresses within the element equalizes the virtual work of the element forces in the nodes. As a consequence, stresses on boundaries of adjacent elements are only equal in an approximate way. The used kinematic equations and constitutive laws normally apply for both thin and thick shells. For thin cylindrical shells the exactness is comparable to the Morley theory of Chap. 9. The solution converges to the exact solution for increasing fineness of the element mesh. Elements are developed in a number of ways. Two approaches are mentioned here.

17.1.1 Method 1

For the first method we refer to shell elements as shown in Fig. 17.2. The element is defined by the shape of the mid plane of the element, the number and position of nodes, the thickness and material properties. We choose a set of local element axes x, y, z. The axes x and y are in the middle plane of the shell surface, the z-axis normal to it. We define in each node degrees of freedom. They are three displacements u_x, u_y, u_z and two rotations φ_x, φ_y. A displacement distribution within the element is chosen for the three displacements u_x, u_y, u_z and the two rotations φ_x, φ_y. This displacement field is an interpolation on the basis of the degrees of freedom in the element nodes, such that continuity across element interfaces is guaranteed. From the displacement field we derive strains and curvatures by application of the kinematic relations. Denote the shell forces (membrane forces, moments and shear forces) by $\mathbf{s}^e$, the membrane strains, curvatures and shear angles by $\mathbf{e}^e$, the vector of element forces in the element nodes by $\mathbf{f}^e$, and the vector of degrees of freedom in the element nodes by $\mathbf{u}^e$. Then, we may write the kinematic relation and constitutive relation as

$$\mathbf{e}^e = \mathbf{B}\,\mathbf{u}^e \tag{17.1}$$

$$\mathbf{s}^e = \mathbf{D}\,\mathbf{e}^e \tag{17.2}$$

respectively. We further may define the relation between $\mathbf{u}^e$ and $\mathbf{f}^e$ with the aid of the element stiffness matrix $\mathbf{K}^e$:

$$\mathbf{K}^e\mathbf{u}^e = \mathbf{f}^e \tag{17.3}$$

Equalizing the virtual work done by shell forces $\mathbf{s}^e$ and by element forces $\mathbf{f}^e$, we obtain

$$\mathbf{K}^e = \iint\limits_{A} \mathbf{B}^T \mathbf{D}\,\mathbf{B}\ dA \tag{17.4}$$

where A is the area of the element. The next step is to assemble the element stiffness matrices to a global stiffness matrix $\mathbf{K}$ and to produce the global load vector $\mathbf{f}$ on the basis of the element load vectors. Taking account of the kinematic boundary conditions, we can compute the displacement vector $\mathbf{u}$ of the shell by solving the matrix equation

$$\mathbf{K}\mathbf{u} = \mathbf{f}. \tag{17.5}$$

Next, we may return to each separate element and compute its vector of element forces from Eq. (17.3) and the element shell forces from Eqs. (17.1) and (17.2):

$$\mathbf{s}^e = \mathbf{D}\mathbf{B}\,\mathbf{u}^e. \tag{17.6}$$

The computation of the element stiffness matrix $\mathbf{K}^e$ in Eq. (17.3) requires integration over the area A of the element. This may sometimes raise difficulties because of shear locking. It is beyond the scope of this chapter to discuss this problem in detail.

Fig. 17.3 Degenerated volume elements with 20 and 15 nodes respectively

17.1.2 Method 2

The second method is equivalent to the first one as for the strategy. Equations (17.1) to (17.6) apply again. The difference is in choosing the element displacement field. One starts from the brick-shaped volume element with 20 nodes for a quadrilateral shell element and from the prismatic volume element with 15 nodes for the triangular shell element. These volume elements are degenerated such that the size in one direction is made small compared to the size in the two other directions as shown in Fig. 17.3. As a rule, volume elements have curved edges. Each node has three degrees of freedom, the displacements u_x, u_y, u_z, so on each short edge, nine degrees of freedom occur. The constraint is introduced that the three nodes on the shortened edges must be on a straight line, and the constitutive equations are slightly adapted. This facilitates reduction of the nine degrees of freedom in three nodes to one five degrees of freedom in one node, the same as in Method 1. Similarly, the nine element forces in the straight line are restyled by summing up to five generalized forces and moments as applicable in Method 1.

17.2 Example of Irregular Shell

17.2.1 Geometry and Mesh

We will demonstrate the use of FEA for the irregular shell of Fig. 17.1. The shell is one out of a row in a circus roof at the beach resort Zandvoort, The Netherlands, and has been designed by Soeters Van Eldonk architects. A global set of axes x, y, z is needed to define the geometry of the shell. For this we refer to the contour plot of the shell geometry in Fig. 17.4. The vertical axis z is upwards. The contour plot reveals that the shell is practically axisymmetric at the top and gradually develops into a rectangular ground plan at lower levels. We will need a local set of axes x_l, y_l, z_l in the mid plane of the shell for the shell forces. That set is different from the set of global axes; the x_l-axis is in the direction of the meridian and the y_l-axis in circumferential direction. The applied program Scia Engineer chooses the local x-axis in circumferential direction and the y-axis meridian, but we changed that in order to harmonize with Chap. 15. The short edge of the shell is 10.700 m and the

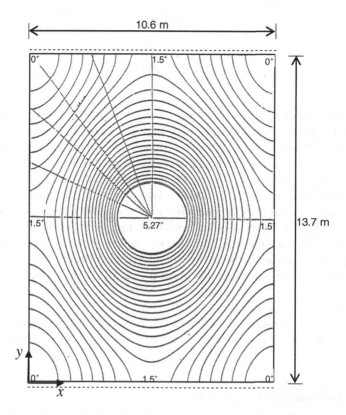

Fig. 17.4 Contour lines of shell surface. Supports at short edges (*dashed lines*)

long one 13.700 m. The thickness is 0.100 m. The radius of the upper opening is 1.500 m. The height is 5.270 m. We used the material properties $E = 20.000$ Nmm^2 and $v = 0.2$. Its own weight is 2.5 kN/m^2.

The shell is clamped at the circular opening at the top. Two lower edges are lines of symmetry, and two simply-supported. In the nodes at the top, all six degrees of freedom (three translations and three rotations) are put to zero. The supported bottom nodes are on the short edges parallel to the x-axis. Here only the vertical translation is put to zero. All other degrees of freedom are free along this edge. At the two bottom edges parallel to the y-axis, we specify that the displacement in x-direction normal to the edge and the rotations about the y- and z-axis are zero. The analysis is done for the load case own weight.

Modern software packages will use mesh generators. As a rule, the user can specify wishes about the average element size and/or indicate zones where refinement is wanted. The choice of the mesh fineness requires some thought about the expected edge disturbance. If we want to properly account for the disturbance at the top, sufficient elements must be chosen over the influence length. The top half of the shells has a shape that is close to a cone. That means that the influence length can be computed with aid of the expression

$$l_i = 2.5\sqrt{r_2 t} \qquad (17.7)$$

where

$$r_2 = \frac{a}{\sin\phi} \qquad (17.8)$$

and a is the radius in the horizontal plane at the top and ϕ the angle between the meridian and the horizontal axis. It holds that $a = 1.667$ m and $\phi = 60°$. Therefore we obtain the influence length $l_i = 1.38$ m. If we want about 10 elements over this length, the average length should not be larger than 0.138 m. We specified an average mesh size of 0.10 m. This results in about 32.000 elements for a complete shell, implying on the order of 200.000 equations. Because the shell has two vertical planes of symmetry, an analysis with about 8.000 elements might do, but we did not profit from this in order to have a check on the modelling.

To generate the total mesh, we started with a quarter of the shell and divided it in four parts. We call lines between the parts *boundaries*. In Fig. 17.1 the lines are visible. We produced separately the mesh for each part. After the mesh for the quarter was complete, we obtained the mesh for the complete shell by copying, pasting and rotating the quarter. Although we did our utmost best to model the shell properly, we did not succeed in making the shell surface continuous at the boundaries as for the tangent plane. There are geometrical imperfections in the model.

17.2.2 Computational Results

The first result we show in Fig. 17.5 is the distribution of the vertical support reactions at the top of the shell. Remember that the lower edge is supported at the two edges parallel to the x-axis. From Fig. 17.5 we notice that the support reaction at the top is not homogeneously distributed over the nodes at the circumference. The highest value occurs at the meridians which run to the supported bottom edges, and the lowest value at the meridians to the not supported bottom edges. The distribution is not fluent, which is due to an inhomogeneous distribution of the nodes over the upper circle. The maximum value is 5.61 kN and the lowest 3.38 kN. The total weight of the shell is 427 kN, of which (rounded off) 360 kN is carried at the top and 67 kN at the two bottom supports. Not less than 84 % of the load is transferred through the upper support.

Figure 17.6 presents a contour plot of the displacement in the z-direction of the global set of axes. The red colour part of the shell close to the upper support is practically not displaced. On the contrary, the dark blue shell parts have maximum displacement. In these regions, the curvatures of the shell are minor and the shell is rather flat. In the top part of the structure shell membrane action is dominant, and the lower part seems to behave like a plate.

Figure 17.7 depicts the distribution of the membrane force n_{xx} along two meridians, ending in the middle of the short supported edge and in the middle of

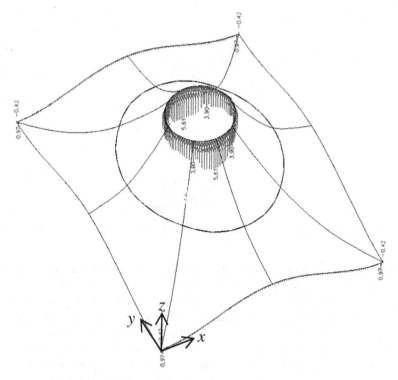

Fig. 17.5 Support reactions at the top of the shell

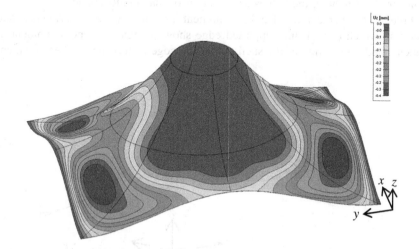

Fig. 17.6 Vertical displacement u_z

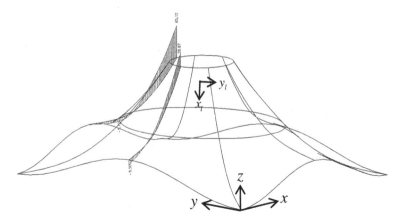

Fig. 17.7 Membrane force n_{xx} (supported edges in global x-direction)

the long edge of symmetry, respectively. Note that the orientation of the shell differs from the previous figure and that the local axis y_l is in the direction of the meridian. The distribution is not homogeneous in circumferential direction, the highest tensile value 65.12 kN/m occurring in the meridian to the supported edge. The membrane force remains tensile over a big part of the shell. Only close to the lower edge, it turns into compression, which becomes zero at the supported edge in correspondence with the boundary condition. The membrane force of 65.12 N/mm occurs over a thickness of 100 mm, so the tensile stress is 0.65 N/mm^2.

Figure 17.8 shows the membrane force n_{yy} in circumferential direction in the same meridians as chosen in the previous figure. The distribution is less plausible than for n_{xx}. Anyhow, the values are an order of magnitude smaller.

Figure 17.9 depicts the bending moment m_{xx} in the two meridians. The meridian which ends at the supported edge shows a significant peak moment at the clamped upper edge of the shell due to the edge disturbance. A zero moment

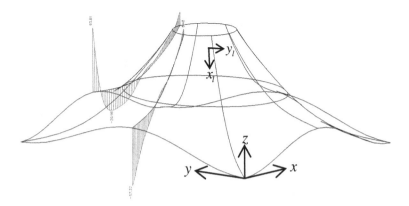

Fig. 17.8 Membrane force n_{yy} (supported edges in global x-direction)

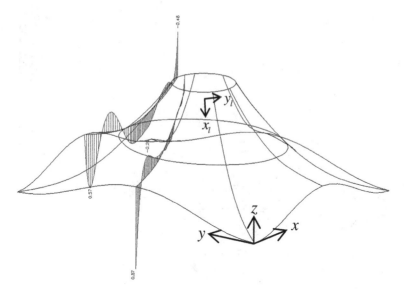

Fig. 17.9 Bending moment m_{xx} (supported edge in global x-direction)

m_{xx} occurs at the supported edge. The distribution of m_{xx} in the other meridian, which ends at the edge of symmetry, shows peak moments at both the top and the bottom. The top peak is due to the edge disturbance. The peak moment at the bottom is not expected. It may be due to the geometric imperfection of the element model.

In Fig. 17.10 we present a full contour plot of the bending moment m_{xx} over a quarter of the shell. In this plot the meridian results of Fig. 17.9 are included. Clearly, the highest values occur in the region that is crossed by the meridian running to the corner of the rectangular ground plan.

17.3 Check of FEA-Results by Theory of Cones

We are able to check the order of magnitude of the analysis results in the top region of the shell. There the shell shape is very close to a cone, and we may calculate an axisymmetric solution as an average of the solution of the real shell. For axisymmetric membrane solutions we use the expressions in Eq. (14.39) and for the bending moment in the disturbed edge zone the flexibility relations of Fig. 15.9. We replace the real shell by a tangent cone, fitting the real shell at the top. The cone is clamped at the top and free at the bottom. The cone's own weight is $p = 2.5 \text{ kN/m}^2$, equal to the real shell's own weight. The length of the replacing cone is chosen such that the weight of the cone yields the same total support reaction at the top as for the real shell, which is 360 kN. Figure 17.11 shows the result for the geometry of the replacing tangent cone. The height is 6.093 m, the top radius $a_t = 1.500$ m and

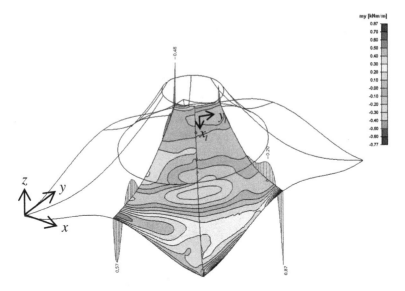

Fig. 17.10 Contour plot of bending moment m_{xx} (supported edges in global x-direction)

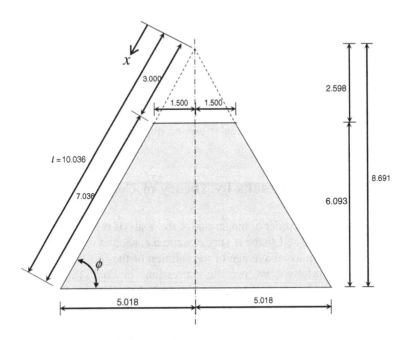

Fig. 17.11 Checking analysis by replacing cone

the bottom radius $a_b = 5.018$ m. The length of the meridian from the apex of the cone to the bottom circle is $l = 10.036$ m, and the angle is $\phi = 60°$. We will first derive the membrane solution and after that the bending moment due to the edge disturbance caused by the stiff ring beam at the top.

17.3.1 Membrane Solution

The solutions of the membrane state are listed in Eq. (14.39):

$$n_{xx} = \frac{pl}{2\sin\phi}\left(\frac{l}{x} - \frac{x}{l}\right) = \frac{2.5 \times 10.036}{2 \times 0.866}\left(\frac{10.036}{x} - \frac{x}{10.036}\right)$$
$$= 14.486\left(\frac{10.036}{x} - \frac{x}{10.036}\right) \text{N/m.} \tag{17.9a}$$

$$n_{\theta\theta} = -px\frac{\cos^2\phi}{\sin\phi} = -2.5\,x\frac{(0.500)^2}{0.866} = -0.722\,x \text{ N/m.} \tag{17.9b}$$

The upper edge of the cone is at $x = 3.000$ m. There we find an axial membrane force $n_{xx} = 44.1$ kN/m, less than the maximum value 65.1 kN/m in the real shell and more than the minimum value 28.87 kN/m. The circumferential membrane force becomes $n_{yy} = 2.17$ kN/m. This compressive membrane force is small compared to the membrane force n_{xx}, but is not yet the complete solution. The edge disturbance has still to be added.

We also can check the membrane force $n_{\theta\theta}$ in circumferential direction. The horizontal displacement at the top circle is zero, hence $\varepsilon_{\theta\theta} = 0$. As a consequence we find $n_{\theta\theta} = \nu n_{xx}$ at the top. Comparison of Figs. 17.7 and 17.8 confirms this. The maximum value is 65.1 kN/m for n_{xx} and 13.0 kN/m (not written in the figure) for $n_{\theta\theta}$.

17.3.2 Bending Moment Due to Edge Disturbance

We can calculate the membrane strain in circumferential direction by

$$\varepsilon = \frac{1}{Et}(-\nu n_{xx} + n_{\theta\theta}) = \frac{1}{20 \times 10^6 \times 0.100}(-0.2 \times 40,58 - 2.17)$$
$$= -5.143 \times 10^{-6} \tag{17.10}$$

The inward horizontal radial displacement of the shell edge is the product of this strain and the radius (1.500 m), so $u_r = 5.143 \times 10^{-6} \times 1.500 = 7.715 \times 10^{-6}$ m. We neglect the small rotation in the membrane state. The displacement u_r must be eliminated by a horizontal force f_r while restricting rotation φ_x of the edge. With reference to case d in Fig. 15.9 we must solve

$$\frac{1}{D_b} \begin{bmatrix} \dfrac{\sin^2 \phi}{2\beta^3} & -\dfrac{\sin \phi}{2\beta^2} \\ -\dfrac{\sin \phi}{2\beta^2} & \dfrac{1}{\beta} \end{bmatrix} \begin{bmatrix} f_r \\ t_x \end{bmatrix} = \begin{bmatrix} 7.715 \times 10^{-6} \\ 0 \end{bmatrix} \qquad (17.11)$$

Herein

$$D_b = \frac{Et^3}{12(1 - v^2)}; \quad \beta^4 = \frac{3(1 - v^2)}{(r_2 t)^2} \qquad (17.12)$$

where $r_2 = a_t / \sin \phi$. With these values we find: $r_2 = 1.5/0.866 = 1.732$, $\beta = 3.130 \text{ m}^{-1}$, $D_b = 1.736 \times 10^3 \text{ kNm}$. Then Eq. (17.11) becomes

$$\begin{bmatrix} 12.23 & -44.19 \\ -44.19 & 319.5 \end{bmatrix} \begin{bmatrix} f_r \\ t_x \end{bmatrix} = \begin{bmatrix} 13.393 \times 10^{-3} \\ 0 \end{bmatrix} \qquad (17.13)$$

from which we derive

$$\begin{aligned} f_r &= 2.19 \text{ kN/m} \\ t_x &= 0.30 \text{ kNm/m} \end{aligned} \qquad (17.14)$$

Because $t_x = -m_{xx}$, we find $m_{xx} = -0.30$ kN. In Fig. 17.9 we see that the maximum value in the real shell is -0.48 kN and the minimum value about -0.17 kN. The axisymmetric cone value -0.30 kN is nicely between these values.

References

1. Love AEH (1927) The mathematical theory of elasticity, 4th edn, Cambridge University Press, Cambridge
2. Reissner E (1966) On the derivation of the theory of elastic shells. In: Proceedings of 11th International Congress on Applied Mechanics. Munich 1964, Springer, Berlin, pp 20–20
3. Wlassow WS (1958) Algemeine Schalentheorie und ihere Anwendung in der Technik, Akademie Verlag, Berlin
4. Morley LSD (1959) An improvement on Donnell's approximation for thin-walled circular cylinders. Q J Mech Appl Math 12:89–99
5. Flügge W (1960) Stresses in shells. Springer, Berlin (corrected reprint 1962)
6. Novoshilov VV (1959) The theory of thin shells. In: Lowe PG (ed) Translated from the Russian (1951), Noordhoff, Groningen
7. Koiter WT (1960) A consistent first approximation in the general theory of thin elastic shells, In: Proceedings of IUTAM symposium on the theory of thin elastic shells, Delft, North Holland, pp 12–33 August 1959
8. Koiter WT (1969) Foundations and basic equations of shell theory: A survey of recent progress, In: Proceedings of IUTAM symposium on theory of thin shells, Copenhagen, Springer, Berlin, pp 93–105 September 1967
9. Donnell LH (1933)Stability of thin-walled tubes under torsion. NACA. Report No. 479
10. Niordson FI (1985) Shell theory (North-Holland series in applied mechanics and mechanics) Elsevier Science, New York
11. Plücher A (1957) Elementare Schalenstatik. Springer, Berlin
12. Rüdiger D, Urban J (1966) Kreiszylinderschalen (Tabellen), B.G. Teubner Verlagsgesellschaft, Leipzig
13. Timoshenko SP, Woinowsky-Krieger S (1959) Theory of plates and shells, 2nd edn, Mc Graw-Hill Book Company Inc., New York
14. Wolmir AS (1962) Biegsame Platen und Schalen, VEB Verlag für Bauwesen, Berlin (DDR)
15. Bouma AL et al (1958) Edge disturbances in axisymmetrically loaded shells of rotation, IBC mededelingen, Inst. TNO for Build. Materials and Build. Structures (In Dutch)
16. Bouma AL et al (1956) The analysis of the stress distribution in circular cylindrical shell roofs according the D.K.J. method with aid of an analysis scheme. IBC mededelingen, Inst. TNO for Build. Materials and Build. Structures (In Dutch)
17. Bouma AL, Van Koten H (1958) The analysis of cylindrical shells (in Dutch), Report No. BI-58-4. TNO-IBBC, Delft
18. Hoefakker JH (2010) Theory review for cylindrical shells and parametric study of chimneys and tanks. Doctoral thesis Delft University of Technology, 2010, Eburon Academic Publishers, Delft
19. Karman T, von Tsien HS (1941) The buckling of thin cylindrical shells under axial compression, J Aeronaut Sci 8:303

J. Blaauwendraad and J. H. Hoefakker, *Structural Shell Analysis*, 295
Solid Mechanics and Its Applications 200, DOI: 10.1007/978-94-007-6701-0,
© Springer Science+Business Media Dordrecht 2014

20. Jenkins RS (1947) Theory and design of cylindrical shell structures, The O.N Arup Group of Consulting Engineers
21. Loof HW(1961) Edge disturbances in a hyppar shell with straight edges, Report 8-61-3-hr-1. Stevin laboratory, Department of Civil Engineering Technical University Delft (In Dutch)
22. Bouma AL (1960) Some applications of the bending theory regarding doubly curved shells. In: Proceedings of symposium on theory of thin shells, Delft, North Holland, Amsterdam, p 202–235, August 1959

Index

J. Blaauwendraad and J. H. Hoefakker, *Structural Shell Analysis*,
Solid Mechanics and Its Applications 200, DOI: 10.1007/978-94-007-6701-0,
© Springer Science+Business Media Dordrecht 2014